Computational Finance
MATLAB® Oriented Modeling

Francesco Cesarone

Computational Finance
MATLAB® Oriented Modeling

LONDON AND NEW YORK

G. Giappichelli Editore

First published 2021
by Routledge
2 Park Square, Milton Park, Abingdon, Oxon OX14 4RN

and by Routledge
52 Vanderbilt Avenue, New York, NY 10017

Routledge is an imprint of the Taylor & Francis Group, an informa business

and by G. Giappichelli Editore
Via Po 21, Torino – Italia

British Library Cataloguing-in-Publication Data
A catalogue record for this book is available from the British Library

Library of Congress Cataloging-in-Publication Data
A catalogue record for this book has been requested

MathWorks Contact Information

The MathWorks, Inc.
3 Apple Hill Drive
Natick, MA, 01760-2098 USA Tel: 508-647-7000
Fax: 508-647-7001
E-mail: info@mathworks.com
Web: https://www.mathworks.com
How to buy: https://www.mathworks.com/store
Find your local office: https://www.mathworks.com/company/worldwide

ISBN: 978-0-367-49293-9 (pbk-Routledge)
ISBN: 978-0-367-49303-5 (hbk-Routledge)
ISBN: 978-88-921-3250-4 (hbk-Giappichelli)
ISBN: 978-1-003-04558-8 (ebk-Routledge)

Typeset in Simoncini Garamond
by G. Giappichelli Editore, Turin, Italy

The manuscript has been subjected to the double blind peer review process prior to publication.

Where there is a will there is a way

Dedicated to my best friend, my father

Contents

Preface

Computational Finance is becoming increasingly important in the financial industry. It is the necessary complement to apply the theoretical models to real-world challenges. Indeed, many models used in practice involve complex mathematical problems, for which an exact or a closed-form solution is not available. Consequently, we need to rely on computational techniques and specific numerical algorithms.

This book aims at combining theoretical concepts and their practical implementation. Furthermore, the numerical solution of models is exploited both to enhance the understanding of some mathematical and statistical notions and to acquire sound programming skills in MATLAB®, which can be useful also in several other programming languages.

Most of the content of this book has been taught for several years at a Master's course in Finance to students with a relatively small background in mathematics, probability and statistics. Hence, the book contains a short description of the fundamental tools needed to address the two main fields of quantitative finance: portfolio selection and derivatives pricing. Both fields are developed here, with a particular emphasis on portfolio selection, where we include recent approaches that have appeared only in the literature.

We develop the ability to place financial models in a computational setting. This supports the understanding of theoretical concepts through their practical application.

Audience

This text is intended for students of Economics, Engineering and Applied Mathematics, and for practitioners who wish to investigate some quantitative procedures in the field of finance. The prerequisites are undergraduate courses in Calculus and Financial Mathematics, even though the readers could acquire these notions on their own with a bit of effort.

Scheme of the book

The book contains more than 100 examples and exercises, together with MAT-LAB codes providing the solution for each of them. The road map of the book is as follows. Chapter 1 is devoted to an introduction to the MATLAB®language and development environment, for programming, numerical calculation and visualization applied to simple calculus and financial problems. Chapter 2 introduces basic concepts in probability and statistics, simplifying as much as possible the discussion. Chapter 3 deals with the main constrained optimization models, mainly focusing on recognizing the type of problems treated, and how to implement and solve them in MATLAB®. In Chapter 4 we address Portfolio Optimization, providing several portfolio selection models mainly based on risk-gain analysis. Chapter 5 presents some probabilistic tools which are used in Chapter 6 for describing three methodologies to price derivatives.

Supplemental material

I created the web page http://host.uniroma3.it/docenti/cesarone/Books.htm containing supplemental materials and updates related to this book. Here you can also find where to download the MATLAB®codes described in the book.

Contact information

In spite of my efforts in drafting and checking the text and the MATLAB®codes, some errors and typos could still remain. For this reason, I strongly appreciate any feedback and suggestions kindly sent to the Email address:
francesco.cesarone@uniroma3.it.

About the author

Francesco Cesarone was born in Rome in 1975. He received a Master's degree in Physics and a Ph.D. degree in Mathematics for Economic and Financial Applications from the Sapienza University of Rome. He initially worked as a researcher in the field of climatology at CNR (National Research Council), then as a PostDoc in finance at the Sapienza University of Rome, and as a consultant for ARPM (Advanced Risk and Portfolio Management, New York, US). Since 2011 he is an Assistant Professor of Computational Finance at the Department of Business Studies of the Roma Tre University. His research interests currently include portfolio selection problems, risk management, risk modeling, and optimal risk decisions, enhanced indexation problems, algorithms for large scale linear, quadratic integer and mixed-integer programming problems, heuristic optimization. He serves as a referee for several scientific journals.

Acknowledgements

First of all, I would like to thank my father Antonio Cesarone who encouraged (or rather "forced") me to write this book. I am grateful to several colleagues from whom I have drawn inspiration on various topics of this project: Flavio Angelini, Massiliano Corradini, Carlo Mottura, and Fabio Tardella. I express my deep gratitude to Jacopo Moretti, Giovanni Lo Russo, and Jacopo Maria Ricci for their practical support during the writing of this book. I would also like to thank Attilio Meucci who taught me how to write a book efficiently, during my collaboration with ARPM. Finally, I am grateful to my family members for their continuous support.

Roma, January 2020,

Francesco Cesarone

What are the rumors about the book?

"Solving problem and decision analysis. Both are carried out, throughout the book, by precise mathematical elements, concepts and functions (Theory), then by fundamental exercises provided with solutions (Practice). Last, but not the least, the book is easy to read and to understand."

Maria Luisa Ceprini, *Professor expert in Welfare System, Research Associate, MIT, Sloan School of Management, USA*

"An essential guide to understand and apply financial key concepts."

Rosella Giacometti, *Professor in Mathematical Methods for Economics and Actuarial and Financial Sciences, Department of Management, Economics and Quantitative Methods, University of Bergamo, Italy*

"The most important knowledge for quantitative analysts is included in the book. Based on the book, Readers can develop their own numerical tools to find optimal investments and hedge the risk."

Young Shin Aaron Kim, *Associate Professor in Finance, College of Business Stony Brook University, New York, USA*

"Praiseworthy in its coverage of both portfolio theory and option pricing at an introductory but rigorous level. A gentle introduction to computational finance for the student with a healthy combination of theory and numerics. It will be a finance student bible."

Mustafa Celebi Pinar, *Professor of Industrial Engineering, Bilkent University, Ankara, Turkey*

"This book is a clever compilation of theory and exercises. It constitutes an excellent source for instructors to organize a course on quantitative finance and for students to get some knowledge on different quantitative procedures in the field of finance."

Justo Puerto, *Professor of Operations Research and Optimization, Facultad de Matemticas, Universidad de Sevilla, Spain*

Part I

Programming techniques for financial calculus

Chapter 1

An introduction to MATLAB®with applications

MATLAB is a numerical computing environment and a programming language. It is one of the most popular programming tools, whose fields of application range from finance to robotics, from computer vision to communications, and much more. One of the main reasons for its spread is due to its peculiar language, which is based on matrix calculus. Developed by MathWorks, MATLAB works on different operating systems, including Microsoft Windows, Mac OS X, and Linux. Chapter 1 is devoted to an introduction to the MATLAB language and development environment including programming, numerical calculation, and visualization applied to simple calculus and financial problems.

This chapter is structured as follows: first, we introduce some basic elements for computing in MATLAB. Then, we describe how to build an ordered sequence of commands in a single file, in order to easily and efficiently implement all instructions to reproduce a model. Furthermore, we address the main statements of MATLAB that allow the user to execute expressions repeatedly, or to impose or check certain conditions. Finally, some exercises on programming and on Financial Mathematics are proposed. We point out that the first part of this chapter is devoted to teaching MATLAB to those who use it for the first time, while the part concerning Financial Mathematics is addressed to readers who already have at least a basic knowledge of this topic.

1.1 MATLAB®basics

In this section, we discuss how to start programming with MATLAB. In particular, we describe the basic syntax (i.e., how commands should be structured, so that MATLAB executes appropriate calculations and operations), the commonly used operators and special characters, and some of the variables, constants, and functions, which are predefined in MATLAB.

1.1.1 Preliminary elements

MATLAB is a useful software both for numerical computations and for graphical visualization. MATLAB stands for Mat*rix* Lab*oratory*, and the arrays are the basic unit of its language. The numerical computations, previously mentioned, aim at the practical implementation of mathematical models. MATLAB has several features: it is an interpreter of commands, a programming language, and provides wide graphics facilities. Furthermore, it is able to effectively communicate with different applications (e.g., Excel).

In addition to the basic platform, MATLAB has a variety of tools, called Toolboxes, for specific applications. For our purposes, the Statistics, the Optimization and the Financial Toolboxes are the most useful ones.

A software strictly linked with MATLAB is Simulink, a graphical programming environment for modeling, simulating and analyzing complex and dynamic systems, but this topic is beyond the aim of this book.

Let us start by running MATLAB. One can generally observe a main window divided in several subwindows, as shown in Fig. 1.1. In this case, we have the Current Folder on the left, the Command Window at the center, and the Workspace and the Command History on the right. The Current Folder (or Directory) represents the place where we are working, the Command History lists, in chronological order, the last commands used, and the Command Window contains the prompt (the blank line that follows the symbol $>>$), where we insert instructions, run MATLAB programs, and so on. Note that the complete path of the Current Folder is generally shown above the Command Window. Fig 1.1 shows an example of the MATLAB main window (corresponding to version R2019b), where, at the top, we can see three sheets: Menu, Plots, and Apps. Generally, these contain elements for managing and elaborating the MATLAB files; for finding bugs; for setting up the parallel programming; for configuring the desktop of MATLAB; for accessing to Help menu. Note that the Help menu and the Mathworks website (http://www.mathworks.it/matlabcentral/) are extremely useful for solving problems, especially for MATLAB beginners.

Variable assignment

Let us consider how to define a variable, for instance, a scalar. As mentioned above, the building block of MATLAB is the array, i.e., an orderly arrangement

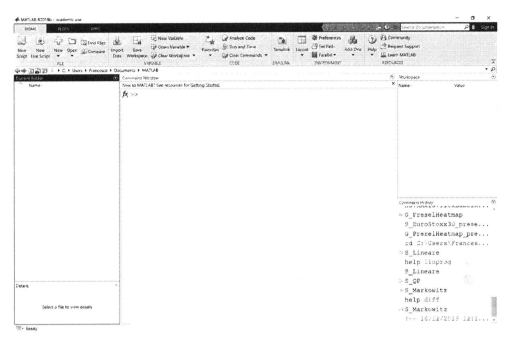

Figure 1.1: Example of the layout of the MATLAB main window.

of cells, usually in rows and columns. If the cells are numbers, we obtain scalars, vectors and matrixes. Otherwise, if the cells are strings, we have an array of strings, and so on. Consider now a scalar, that is a 1×1 matrix. To define a new variable equal to 3.2, one can write on the prompt the following instruction

>> x=3.2;

where x is the name of the variable, and 3.2 is its assigned value. The name of a variable must not exceed 31 continuous characters and must not contain mathematical operators (-,+,*,=), apostrophe, punctuation, slash, or backslash. If one writes a scalar on the prompt (>>), a default variable is generated and it is indicated by **ans**. For instance, one can type

>> 1.67;

Note that adding the semicolon (;) at the end of an instruction avoids showing the output on the Command Window and decreases the running time of an algorithm. Conversely, in order to visualize the assigned variables, one can run the name of a variable (for example, x or **ans**) on the prompt; otherwise, it is possible to directly see the variable in the Workspace subwindow (see Fig 1.1).

13

There are different options to visualize numbers associated to the assigned variables. Below, various formats are illustrated:

FORMAT (SHORT)	four digits after the comma	2.7000
FORMAT LONG	fifteen digits after the comma	2.700000000000000
FORMAT SHORT E	exponential short form	2.7000e+000
FORMAT LONG E	exponential long form	2.700000000000000e+000
FORMAT SHORT G	best representation with four digits	2.7
FORMAT LONG G	best representation with fifteen digits	2.7
FORMAT BANK	numbers expressed in euro and euro cents	2.70

Furthermore, MATLAB provides some predefined variables listed in the following table:

Variable	Meaning
ans	Value of the last operation, to which a name is not assigned
i	Imaginary unit
pi	Approximation of π
eps	Machine accuracy
realmax	Maximum positive machine number
realmin	Minimum positive machine number
Inf	inf (namely a number greater than realmax)
NaN	Not a Number (e.g., $\frac{0}{0}$)
computer	the type of your computer
version	MATLAB version

Workspace

The Workspace is represented by a subwindow and is the place where assigned variables are stored. It can show several details of the assigned variables, for instance:

Name	name of the variable
Value	assigned value
Size	array dimension (rows $\times$ column)
Bytes	allocated memory
Class	variable types

The MATLAB variable types are *double, cell, sparse, char, struct,* and *uint8*. By default, MATLAB works with variables in double precision. A cell array is an array whose elements are, in turn, arrays with different possible dimensions.

A sparse matrix is generally a matrix with most of the elements equal to zero, where, therefore, only the elements different from zero are represented. For more details, see the MATLAB Help. Another feature of the Workspace is that it allow us to open a window similar to an Excel spreadsheet by double-clicking on the variable.

Note that MATLAB is case sensitive. Thus, the variables x and X are considered different variables.

Arithmetic operations

The basic calculation operations are the arithmetic ones: sum (+), subtraction (−), multiplication (∗), division(/), and power (^). For instance, if we have to calculate

$$x = \frac{6 + 9^3 - 8/2}{2 * (2 + 1)^4}$$

we can write the following instruction on the command prompt:

>> x=(6+9^3-8/2)/(2*(2+1)^4);

In the MATLAB environment, the arithmetic operations work in the following order:

Order	Operation	Example
I	Brackets	(3+4)*4=7*4
II	Power	3^2+2=9+2
III	* /, from left to right	2*3/5=6/5
IV	+ −, from left to right	8-5+3=3+3

However, one can change the order of the operations by appropriately applying parentheses (). If an instruction consists in a long expression, one can divide it in two or more lines by writing three consecutive points, as shown in the following example:

>> x=1/2+(9*3)^3-...
3+12;

These instructions compute the expression $x = \frac{1}{2} + (9 * 3)^3 - 3 + 12$.

1.1.2 Vectors and matrices

The building block of MATLAB is the array. An array is an ordered succession of memory locations that contains a collection of elements of the same type. As specified in Section 1.1.1, an array can be made by six types of variables: char, double, `sparse`, `cell`, `uint8`, `struct`. Consider now an array a (1×5), namely 1 row and 5 columns (a row vector), that contains integer numbers from 1 to 5. As shown below, we can create a in several ways:

```
>> a=[1 2 3 4 5];
>> a=[1,2,3,4,5];
>> a=1:5
```

```
a  =

    1   2   3   4   5
```

To generate an array b (5×1), namely 5 rows and 1 column (a column vector), that contains integer numbers from 1 to 5, for instance, we can write:

```
>> b=[1;2;3;4;5]
```

```
b  =

    1
    2
    3
    4
    5
```

Note that the space or the comma (,) separate elements on the same row, while the semicolon (;) separates elements on the same column. To define an array c (2×5), i.e., a matrix with 2 rows and 5 columns, we can opportunely combine the space (or the comma) and the semicolon as follows:

```
>> c=[15 2 4 -1 9; 1 5 -7 0 7]
```

```
c  =

    15    2    4   -1    9
     1    5   -7    0    7
```

Detecting the dimensions of a matrix becomes very useful when the number of its elements is too large to visualize them directly. There are two built-in functions that allow for the identification of array dimensions, namely `size` and `length`. Applying the function `size` to a generic matrix $m \times n$ generates the number of rows (m) and columns (n) as outputs. In the case of the matrix c just defined, we have

```
>> size(c)

ans =

    2   5
```

The function `length` is usually used to detect the number of elements in a vector. However, applying the function `length` to a matrix gives the greater number between m and n as output. In the previous example, we obtain

```
>> length(c)

ans =

    5
```

Then, we have that `length(c)=max(size(c))`.

Another important issue to address is how to access and manipulate the elements of an array. For instance, given the matrix c, we can detect the first element of the second row in the following way:

```
>> c(2,1)

ans = 1
```

A similar procedure can be used for a vector. For example, given the vector b, to identify its second element, we can write

```
>> b(2)

ans = 2
```

Furthermore, equating a specific element to a scalar replaces that element with the newly assigned scalar. For instance, continuing the previous example, we have

```
>> b(2)=6

b =

    1
    6
    3
    4
    5
```

17

To modify an element or a set of elements in a matrix, we have to suitably specify two indices. For instance, considering the matrix c, to modify the element $c_{1,3}$, we can write

```
>> c(1,3)=21

c  =

      15    2    21   -1    9
       1    5    -7    0    7
```

Note that running the command `d(3,4)=3` without predefining the matrix `d` generates a 3×4 matrix with all elements equal to zero, except for that of position $(3, 4)$:

```
>> d(3,4)=7

d  =

     0   0   0   0
     0   0   0   0
     0   0   0   7
```

1.1.3 Basic linear algebra operations

As mentioned above, the natural environment of MATLAB is the matrix calculus. Its standard operations are the following:

  + → element-by-element sum of matrices (or vectors)

  - → element-by-element difference of matrices (or vectors)

  * → row-by-column product of matrices (or vectors)

Note that, for the sum and the difference, i.e., for element-by-element operations, the matrices (or the vectors) involved must have the same dimensions. While for the row-by-column product the number of columns of the first matrix must be equal to the number of rows of the second matrix. For instance, given the vectors a and b previously introduced, and the column vector `a_=[1;2;3;4;5]`, we can implement the following operations

```
>> a_+b          >> a_-b      >> a*b

ans  =           ans  =       ans  =
         2                0            63
         8               -4
         6                0
         8                0
        10                0
```

Note that, if we compute the difference between the vectors a e b, we obtain the following output:

```
>> a-b

Error using ==> -
Matrix dimensions must agree.
```

Similarly, we can apply the basic linear algebra operations among matrices. For instance, consider the following matrices

$$e = \begin{pmatrix} 2 & 1 & 2 \\ 3 & 0 & 1 \end{pmatrix} \quad \text{and} \quad f = \begin{pmatrix} 2 & 1 \\ 1 & 0 \\ 0 & 1 \end{pmatrix},$$

and calculate their product:

```
>> g = e*f

g =
        5        4
        6        4
```

This result is a consequence of the matrix multiplication rule, which, in terms of dimensions, works as follows: $(2 \times 3)(3 \times 2) = (2 \times 2)$.

Other standard operations are the *transposition* and the *inversion* of matrices. Below we show these operations by means of two examples. Firstly, we transpose the matrix e

```
>> e'

ans =
        2 3
        1 0
        2 1
```

Note that if we consider

```
>> f*e'
```

we obtain the following output

```
???  Error using ==> mtimes
Inner matrix dimensions must agree.
```

Indeed, in this case, the number of columns of the first matrix is not equal to the number of rows of the second matrix.

Furthermore, we can invert the non-singular square matrix g defined above as follows:

```
>> g_inv=inv(g)

g_inv =
      -1.0000   1.0000
       1.5000  -1.2500
```

Obviously, it must hold that $g^{-1} \cdot g = I$, where I is the identity matrix

```
>> g_inv*g

ans =
       1.0000   0.0000
       0.0000   1.0000
```

For further information about matrix inverse, see the MATLAB Help (`mldivide` \).

Example 1 (Linear equations system) *Let us consider the following system of linear equations*

$$\begin{cases} 3x_1 + 4x_2 = 25 \\ x_1 - 6x_2 = 50 \end{cases}$$

In matrix notation, we can write that $Ax = b$, where $A = \begin{pmatrix} 3 & 4 \\ 1 & -6 \end{pmatrix}$, $x = \begin{pmatrix} x_1 \\ x_2 \end{pmatrix}$, and $b = \begin{pmatrix} 25 \\ 50 \end{pmatrix}$. Since A is an invertible matrix (i.e., $\det(A) = -22 \neq 0$), we have $x = A^{-1}b$. Using MATLAB, we can write:

```
>> A = [3 4; 1 -6];
>> b = [25;50];
>> x = A\b


x =


    15.9091
    -5.6818
```

Note that the product between the inverse of A (A^{-1}) and b can be obtained with the backslash $(\backslash)$ command.

1.1.4 Element-by-element multiplication and division

In many cases, it could be useful to perform element-by-element power, multiplication, and division operations among matrices. MATLAB provides these operations by adding a dot (.) before the commands ^, * or / (i.e., .^, .*, ./). For instance, to define a new vector whose elements are squares of the elements of the vector b (modified at the end of the previous section), we can write

```
>> b.^2

ans =

    1
    36
    9
    16
    25
```

Observe that the expression b^2 does not make sense, since it corresponds to the matrix product $(5\times1)(5\times1)$.

Example 2 (Element-by-element operations) *Compute the element-by-element product and division between the vectors* a_2=[2,2,2,2,2] *and* b_2=[4,8,16,2,10]. *On the command prompt, we can type:*

```
>> a_2.*b_2

ans =

    8   16   32   4   20
```

and
```
>> a_2./ b_2

ans =

    0.5000   0.2500   0.1250   1.0000   0.2000
```

Note that the involved vectors (or matrices) must have the same dimensions.

1.1.5   Colon (:) operator

This command is frequently adopted for several purposes. For instance, it can be used to create vectors with equally spaced elements, such as a set of numbers that identifies the index $k = 1, 2, \ldots, n$, by writing $>>$ k = 1:n. As a consequence, the colon operator (:) can be exploited to select rows or columns of an array. For instance, considering the matrix c of Section 1.1.2, we can select the first two rows and columns of c as follows

```
>> c(1:2,1:2)

c =

    15   2
     1   5
```

or we can extract the third column as follows:

```
>> c(:,3)

ans =

    21
    -7
```

More generally, the command Vector=Start:Step:End generates a vector with equally spaced elements between the number Start and the number End. For example, to create a vector of odd numbers that starts from 1 up to 15, we can write:

```
>> h=1:2:15

h =

    1 3 5 7 9 11 13 15
```

By default, Step (when it is omitted) is equal to 1. If Step is positive, then the value of the elements of Vector increases. Otherwise, if Step is negative, then the elements of Vector decrease as shown in the following example:

```
>> k=12:-2:4

k =

    12 10 8 6 4
```

Note that the colon operator (:) will be extensively used when addressing the loop scheme in MATLAB (see Section 1.3.2).

When Step is not an integer, it could be worth using the *built-in* function linspace, where the inputs are the first and the last element of the vector, and the number of components of the vector. For example, we can write

```
>> l=0; m=1; n=8;
>> x=linspace(l,m,n)
x=
    Columns 1 through 4
    0 0.1429 0.2857 0.4286
    Columns 5 through 8
    0.5714 0.7143 0.8571 1.0000
```

It generates a vector x with equally spaced elements with a step equal to $\dfrac{m-l}{n-1}$. More in detail, the elements of x are

$$x_i = l + (i-1)\frac{m-l}{n-1} \quad \text{with} \quad i = 1, ..., n;$$

where x_i indicates the *i-th* element of x.

23

1.1.6 Predefined and user-defined functions

MATLAB provides a number of mathematical functions. A non-exhaustive list of predefined function is reported in the following table

Function	Meaning
`sin, cos, tan`	sine, cosine, tangent
`asin, acos, atan`	arcsine, arccosine, arctangent
`exp`	exponential
`sinh, cosh`	hyperbolic sine, hyperbolic cosine
`tanh`	hyperbolic tangent
`log, log2, log10`	logarithm to the base e, 2, and 10
`sqrt`	square root
`abs`	absolute value
`sign`	sign function

The scheme of a predefined functions is `Output=function(Input)`, where `Output` and `Input` can be, in general, matrices. For instance, if we apply the function $\sin$ to the matrix c of Section 1.1.2, we have

```
>> y=sin(c)


y  =


      0.6503    0.9093    0.8367   -0.8415    0.4121
      0.8415   -0.9589   -0.6570        0    0.6570
```

inline Function

MATLAB also gives the possibility to create any function with one or more independent variables by means of the `inline` function. As specified in the MATLAB Help, to build a generic function `anyf`, we can use the following syntax:

```
>> anyf=inline('expr','arg1','arg2',...,'argn');
```

where the string `expr` provides the mathematical expression of the function considered, and `arg1,arg2,...,argn` indicate the independent variables of this function. Below we report two examples of functions with one (on the left) and two (on the right) independent variables:

```
>> f=inline('x.^2.*tanh(x)')      >> g=inline('sqrt(x.^2+y.^2)','x','y')
f=                                 g=
    Inline function:                   Inline function:
    f(x)=x.^2.*tanh(x)                 g(x,y)=sqrt(x.^2+y.^2)
```

24

Note that both the expressions and the arguments must be included within apostrophes. Therefore, it is possible to evaluate the two previous functions w.r.t. any fixed points. For instance, we can type

```
>> x=2.5;          >> x1=1.5; x2=3.7;
>> y=f(x)          >> z=g(x1,x2)


y =                z =


        6.1663             3.9925
```

Furthermore, if the mathematical expression `expr` is properly written, then the inputs `arg1,arg2,...,argn` of the generic function `anyf` can also be matrices. For instance, referring to the function g, we have

```
>> x1=[0 1]; y1=[1 2];
>> g(x1,y1)
ans =
        1 2.2361
```

Observe that the names of the inputs do not have to be necessarily the same of those of the variables `arg1,arg2,...,argn` used to define the `inline` function.

Anonymous Function

An alternative to `inline` is the Anonymous Function, that will substitute the former in a future release of MATLAB. However, the syntax to define an Anonymous Function is similar to that of `inline`. For instance, to create an anonymous function that calculates the quartic root of a variable, we can write

```
>> q_root=@(x) x.^(1/4)
```

where `q_root` is the name of the function. The content in brackets following @ represents the independent variables of the mathematical expression of `q_root`. Then, we can calculate the quartic root of $x = 81$ as follows

```
>> x=81;
>> y=q_root(x)
y=
        3
```

Note that an Anonymous Function can include other anonymous functions. However, for further details, see the MATLAB Help.

1.2 M-file: Scripts and Functions

The main programming tools of MATLAB are represented by Scripts and Functions. They can be written by the MATLAB Editor, which provides several facilities in the programming review. Script and Function are saved as .m files, a specific extension of MATLAB (also called M-file). In a nutshell, in an M-file we can write all the instructions to implement a model and run Script or Function on the prompt. For convenience, we summarize the main characteristics and differences between Script and Function in the following table

Script	Function
works on variables in the Workspace	*the interior variables are local*
does not accept input variables	*can accept input variables*
does not have output variables	*can have output variables*
useful for running a number of instructions	*useful for solving recurrent applications*

For instance, consider the following M-files:

S_HypRightTri.m	**F_HypRightTri.m**
`a= 3;`	`function [c] = F_HypRightTri(a,b)`
`b= 4;`	`c=sqrt(a.^2+b.^2)`
`c=sqrt(a.^2+b.^2)`	`end`

In this case, both Script (left side) and Function (right side) have the same objective, the calculation of the hypotenuse of a right triangle. We can run these M-files on the Command Window as follows:

`>> S_HypRightTri`	`>>c= F_HypRightTri(3,4)`
`c=`	`c=`
`    5`	`    5`

Note that a Function M-file must be saved with the same name as its reference function.

Since the construction of an M-file is a key issue in MATLAB programming, we analyze, in detail, the resolution of a linear system, where the incomplete matrix A is a Hilbert matrix, an example of an ill-conditioned matrix. Then, in

Example 3 the aim is to solve the linear equation system $Ax = b$ and to compute the absolute and relative errors of the obtained solution (due to the instability of the solution when A is an ill-conditioned matrix).

Example 3 (Ill-conditioned matrix: Script) *Open a Script on the Current Folder and call it, e.g.,* S_IllCond.m. *Write the following lines of code*

```
clear all                % delete all variables
                         % in the Workspace
close all                % close all figures
n=30;
A=hilb(n);               % define an n-by-n
                         % Hilbert matrix
x=[1:n]';                % x is a column vector of
                         % n elements
b=A*x;                   % define the vector b
x1=A\b;                  % solve the linear system Ax=b
                         % note that the back-slash
                         % stands for A⁻¹ * b
abs_error=norm(x-x1)     % calculate the absolute error
rel_error=abs_error/norm(x)  % calculate the relative error.
```

On the left of the code, we report the syntax used to solve our problem, while, on the right, some comments are provided. Comments are useful in managing long Scripts (think about a code of more than 300 lines) and can be inserted by typing % before the desired note. Thus running the nae of the Script on the command prompt, MATLAB automatically executes each code line of the Script in chronological order. In addition, if the Script presents some errors, then it is automatically highlighted on the Command Window with indication of the bugged line's location.

The other important M-file is the Function, which is used to implement a block of commands frequently applied in solving a problem. For convenience, we solve the same problem of Example 3.

Example 4 (Ill-conditioned matrix: Function)
Open the editor, and save an M-file with the same name of the reference function, namely F_IllCond.m, *and, therefore, type the following syntax:*

```
      function [abs_error,rel_error] = F_IllCond(n)    % define inputs
                                                       % and outputs
      % Solve a linear equations system               % text visualized
                                                       % in the Help
      A=hilb(n);                                       % n-by-n Hilbert
                                                       % matrix
      x=[1:n]';                                        % a column vector
                                                       % of n elements
      b=A*x;                                           % define vector b
      x1=A\b;                                          % namely A^{-1}·b
      abs_error=norm(x-x1);                            % absolute error
      rel_error=abs_error/norm(x);                     % relative error
```

By means of the Function F_IllCond, *it is possible to compute the absolute and the relative errors for any dimension n of the square matrix A. Given the* 4×4 *and* 7×7 *Hilbert matrices, one can write the following instructions on the prompt*

>> [abs_err4,rel_err4]=F_IllCond(4);
>> [abs_err7,rel_err7]=F_IllCond(7);

Note that the variables in a Function (as for inline *and for an Anonymous function, see Section 1.1.6) are local, and then the names of input and output variables can be different from their names used in the code.*

Let us examine the elements of a Function separately. The first line generally contains the syntax definition for the new function as follows:

function [outputs]=NameOfTheFunction(inputs).

Note that inputs and outputs can be one or more arrays of different types (numbers, strings, etc.). The following lines could contain a description of the Function, which is possible to visualize by typing help NameOfTheFunction on the prompt. After that, there is the body of the Function, where all the variables are local.

In the following table, we synthetically list all possible contents in a Script, or in a Function:

Content of a Script/Function
Commands to load and save data
Assignments
Comments
Calculations
Calls of other defined functions
`for` and `while` loops
`if`, `elseif`, `else` schemes
Commands to construct graphs
etc.

1.3 Programming fundamentals

In this section, we deal with the main basic schemes of programming in MAT-LAB.

1.3.1 `if`, `else`, and `elseif` construct

In some cases, it could be useful to run statements only when specific conditions are verified. The command `if` evaluates an expression that represents a condition. If that condition is true, namely the real part of that expression has all non-zero elements as outputs, then it executes a group of statements. The general form of the `if` scheme is

```
if  expression
statements
elseif  expression
statements
else  expression
statements
end
```

where the `elseif` and `else` lines are optional. Indeed, it is possible to construct a scheme of conditional statements by means of the following short form

```
if  expression
statements
end
```

To formulate one or more conditions, *relational* and *logical* operators can be exploited. Below we provide a list of all *relational* operators available.

Relational operator	Description
<	smaller than
<=	smaller than or equal to
>	greater than
>=	greater than or equal to
==	equal to
~=	different from

The relational operators can be used to verify a condition between arrays with the same dimensions. More precisely, if an expression is true, i.e., the corresponding output is non-empty, and has all non-zero elements, then the statements are executed. Otherwise, the expression is false and the statements are not performed.

Example 5 (Relational operators) *Consider the following two 3×3 matrices, named A and B*

```
>>  A=[11 2 3; 5 10 3; 2 3 2];
>>  B=[2 7 6; 9 10 7; 2 3 2];
```

and check which elements are equal. Thus, we can type the following expression on the prompt

```
>>  A==B
ans=
         0   0   0
         0   1   0
         1   1   1
```

Note that, in the output, 1 indicates that the relation is true, while 0 indicates that the relation is false.

The *logical* operators in MATLAB are summarized in the following table:

Operator	Description
&	and
|	or
~	not

30

As for the *relational* operators, we show how the *logical* operators work by means of the following example.

Example 6 (Logical operators) *Consider an expression generated from two conditions, that, in turn, are related by the operator &. The expression is true if and only if both conditions are true. In numerical terms, the expression is true if both conditions are different from 0.*
For instance, consider the following row vectors

```
>> u=[1 1 0 1 1 0];
>> v=[0 1 1 0 1 0];
```

and write the following statement on the prompt

```
>> u & v
ans=
     0   1   0   0   1   0
```

Thus, if both the elements of u and v are equal to 1 (i.e., the two conditions are true), then the output is equal to one, i.e., the expression is true. Otherwise, the output is equal to 0, i.e., the expression is false.
On the contrary, an expression with the operator | is true if and only if at least one of the conditions is true. In numerical terms, the expression is false if both the operands are equal to zero. Otherwise, the expression is true, i.e., the output is equal to 1. Thus, given the vectors u and v previously defined, we can type

```
>> u | v
ans=
     1   1   1   1   1   0
```

Finally, an expression with the operator ~ consists in the negation of a condition, i.e., the logical complement of that condition. In numerical terms, for each element of an input different from 0, the corresponding element of the output is equal to 0, i.e., it is false, and viceversa. For example we have

```
>> ~ u
ans=
     0   0   1   0   0   1
```

To better understand how the `if`, `else`, and `elseif` block works, try to replicate the following example.

> **Example 7 (if block)** *Create two Scripts where* relational *and* logical *operators are used to execute statements, and run them on the prompt.*
>
> | S_IfBlock_ex1.m | S_IfBlock_ex2.m |
> |---|---|
> | ```x=3;``` | ```x=3;``` |
> | ```if x > 1 | x < 2``` | ```if x > 1 & x < 2``` |
> | ``` disp('well done')``` | ``` disp('well done')``` |
> | ``` y=3*x``` | ``` y=3*x``` |
> | ```end``` | ```end``` |
>
> *Examine in detail the outputs of each code line of these Scripts.*

However, for more details on `if`, `else`, and `elseif` statements, see the Help of MATLAB.

1.3.2 for loops

A `for` loop is a block of commands used to repeat a group of instructions a specified number of times. The general form of a `for` loop scheme is

```
for Index = Start : Step : End
       Statements
   end
```

Note that, if `Step` is not indicated, by default, it is set equal to 1. Furthermore, if `Step` is > 0, then the loop terminates when `Index` becomes greater than `End`. Conversely, if `Step` is < 0, then the loop terminates when `Index` becomes smaller than `End` (see Section 1.1.5).

Let us consider the following partial sum

$$\sum_{i=1}^{n} b_i = b_1 + b_2 + ... + b_n \qquad (1.1)$$

and let us set up how to compute this sum using a `for` loop. Recall that the sum is a binary operation, namely operates between two addends. Thus, to obtain the above partial sum, one can perform a first sum between $b_1 + b_2$, and therefore add b_3 to this first partial result, and so on. To formulate Expression (1.1) in MATLAB with a `for` loop, one can initialize a variable, called "accumulating variable", to 0, and then store all sequences of partial sums. This procedure is applied to the following example.

Example 8 (for loop scheme) *Given a vector* $b = [5, -1, 0, 40, 4, 2, 3]$, *create a Script where all its elements are summed by a* `for` *loop. Below a possible Script is suggested.*

```
b=[5 -1 0 40 4 2 3];
S=0;
n=length(b);
for i=1:n
    S=S+b(i);
end
```

It is worth highlighting the remarkable importance of the `for` loops, that, as you shall see, will be widely used throughout this book. For more details on `for` loops, see the MATLAB Help and some exercises proposed in Section 1.5.

1.3.3   `while` loops

A `while` loop repeatedly executes a group of statements as long as its expression remains true. The general form of a `while` loop is

```
while Expression
       Statements
   end
```

Note that `Expression` is evaluated before the `Statements` are executed for the first time. If `Expression` is true, then the loop works, and `Expression` and `Statements` are evaluated for each iteration. The loop stops when `Expression` becomes false. This scheme is applied to the following example.

Example 9 (while block) *Create a Script to sum the first n natural numbers with* $n = 20$, *by using a* `while` *loop. Below a possible Script is suggested.*

```
i=0;
S=0;
n=20;
while i < n
    i=i+1;
    S=S+i;
end
```

For more details on `while` loops, see the MATLAB Help and some exercises proposed in Section 1.5.

33

1.4 MATLAB®graphics

One of the most useful tools of MATLAB is definitely to create plots. There are different types of graphs, each achievable by a specific built-in function. Below, we give only an introductory description of the main commands of the MATLAB graphics environment. However, further details can be found in many of the exercises contained in this book.

The most commonly used function to perform a figure is `plot`. As shown in the Help, this function is used to plot the elements of a vector (or of a matrix), and then it is possible to feed `plot` on one or multiple objects. However, we mention only few main cases. Let $x_1, x_2, \ldots, x_n, y_1, y_2, \ldots, y_n$ denote generic vectors with the same dimensions, and let X denote a generic matrix. In this case, the following syntax can be used to graph one or more lines on a plane.

- `plot(y1)` graphs the elements of the vector y_1 on the vertical axis versus their indices (from 1 to `length(y1)`) on the horizontal axis.

- `plot(x1,y1)` graphs the elements of the vector x_1 on the horizontal axis versus those of the vector y_1 on the the vertical axis.

- `plot(x1,y1,x2,y2,...,xn,yn)` plots n lines, where the i-th line corresponds to the graph generated by pair vectors x_i, y_i.

- `plot(X)` plots the columns of the matrix X on the vertical axis versus the row indices of X on the horizontal axis.

In the following examples, we propose simple exercises involving the use of the `plot` command for creating graphs and the use of the `print` command to save them.

> **Example 10 (My first graph)** *Create a Script to plot the function $y = \sin(x)$ in the interval $[0; 2\pi]$, and save it in the .jpeg format. Below a possible Script is suggested.*

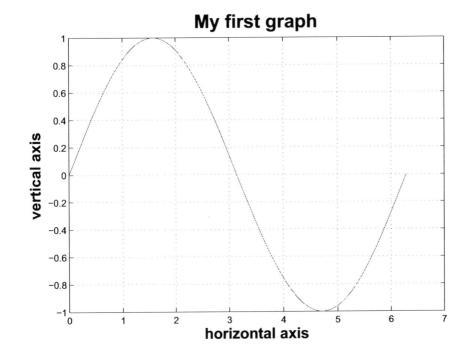

Figure 1.2: Example of plot in MATLAB.

```
clear all            % clear the workspace
close all            % close all figures
clc                  % clear the Command Window
% blank line
x=0:pi/100:2*pi;     % starting point-step-ending point
y=sin(x);            % compute the elements of vector y
plot(x,y)            % plot x (hor axis) versus y (ver axis)
grid                 % add grid lines on the plot
% title
title('My first graph','FontSize',20,'FontWeight','bold')
% x-axis label
xlabel('horizontal axis','FontSize',16,'FontWeight','bold')
% y-axis label
ylabel('vertical axis','FontSize',16,'FontWeight','bold')
% blank line
print -djpeg first_graph % save the graph as .jpeg file
```

Running the above Script will generate a plot as in Fig. 1.2.

Example 11 (Multiple lines graph) *In the interval* $[-3; 3]$*, plot three functions:* $y_1 = e^x$*,* $y_2 = e^{-x}$*,* $y_3 = 5\sin(3x)$*. Furthermore, save it as a .tiff file. Below, a possible Script is suggested.*

```
clear all          % clear the workspace
close all          % close all figures
clc                % clear the Command Window
% blank line
x=-3:0.1:3;        % start-step-end
y1=exp(x);
y2=exp(-x);
y3=5*sin(3*x);
plot(x,y1, x,y2, x,y3)    % plot three lines
print -dtiff second_graph % save the graph
                   % as .tiff file
```

1.5 Preliminary exercises on programming

In this section, we propose several exercises on programming and on Financial Mathematics, where you could apply the knowledge acquired in the previous sections of this chapter. Some of these exercises, although requiring a basic knowledge of finance, focus on the use of programming techniques. Therefore, they are appropriate for all MATLAB novices, regardless of the field of origin.

Exercise 12 (Calculations) *Calculate the following expressions:*

$$6 \cdot 3; \quad \frac{4}{7} \cdot 7 + 13; \quad 2 + \frac{7}{4 \cdot 5}; \quad \frac{4 + 1/3}{9}; \quad \frac{6^3}{5}; \quad (3-1)^{\frac{8}{3}}; \quad 4 + \frac{3}{2^{(2-1)}}.$$

Sol.: See Script S_Calculations.

Exercise 13 (Calculations 2) *Solve the following items.*

1. *Define the variables* $x_1 = \sin(\pi)$*,* $x_2 = 2.9$*,* $x_3 = \sqrt{13}$*,* $x_4 = \ln(3)$*,* $x_5 = e^2$*,* $x_6 = \sqrt[3]{125}$*.*

2. *Delete* x_2 *and* x_6 *from your Workspace.*

3. *Save the other variables in a Workspace (named* `trial.mat`*).*

4. *Delete your temporary Workspace.*

36

5. Load the workspace of point 3, and define a new variable
$$y = \frac{2\sin(\pi) + \sqrt{13}}{(\ln(3))^2}.$$

6. Save the variable y of the previous point in the same workspace as point 3 (adding it to the other variables), and in a different workspace.

Sol.: See Script S_Calculations_2.

Exercise 14 (Manipulation of matrices) *Define a 16×16 matrix A with the command* A=rand(16). *Then, construct a matrix B with the odd columns of A, and a 7×7 matrix C selecting a principal submatrix of A (the first 7 rows and columns).*

Sol.: See Script S_ManipMatr.

Exercise 15 (Save and load) *Generate and save some $m \times n$ matrices (where m and n are arbitrary) as follows:*

1. *Create a matrix A in a .txt file, and load it by using the built-in function* load *or* importdata.

2. *Create a matrix B in a .xls file, and import it in your Workspace (see function* xlsread*).*

3. *Save all the matrices in a Workspace.*

4. *Save the matrix A in a .xls file (see function* xlswrite*).*

Sol.: See Script S_Save_Load_Arrays.

Exercise 16 (Matrix construction) *Create the following matrices.*

1. *A 4×5 matrix, named O, where all the elements are equal to 1, and a 4×5 matrix, named Z, where all elements are equal to 0 (see functions* ones *and* zeros, *respectively).*

2. *A 3×7 matrix with all elements equal to 2.*

3. *A column vector of 100 elements generated by a standard normal random variable (see* randn*).*

4. *A 5×8 matrix randomly generated by a standard uniform random variable (see* `rand`*).*

5. *Save the matrices of the points 1, 2, and 3 in a Workspace.*

6. *Add the matrix of the point 4 to this Workspace.*

7. *Create another Workspace containing only the matrices of the points 1, 2, and 3.*

Sol.: See Script `S_Matrix_construction`.

Exercise 17 (Portfolio of bonds) *Consider a time vector expressed in years* $t = [0.5, 1, 1.5, 2, 2.5, 3, 3.5, 4, 4.5, 5]$*, and the following five coupon bonds with face value* $K = 100€$*:*

1. *a Zero Coupon Bond (ZCB) with 6 months to maturity;*

2. *a Zero Coupon Bond (ZCB) with 1 year to maturity;*

3. *a Coupon Bond with 2 years to maturity, and with semi-annual coupon payments of 0.5 €;*

4. *a Coupon Bond with 3 years to maturity, and with semi-annual coupon payments of 0.75 €;*

5. *a Coupon Bond with 5 years to maturity, and with semi-annual coupon payments of 1.05 €.*

Construct a matrix C *(5×10), where each row represents the flow of the monetary amounts of a bond and each column represents the monetary amounts of the five bonds for each due time. Save the matrix* C *in a Workspace, named* `CB.mat`*.*

Furthermore, create a row vector $\alpha = [-50, -50, 100, 200, 300]$*, which represents the allocation of these bonds in a portfolio. Each component* α_i *represents the units of each bond in the portfolio. Consider now the matrix product (row times column)* $\alpha \cdot C$*. Check its dimension, and think about this output. Create an array* $C(\alpha)$ *(5×10), where each row is the flow of monetary amounts of units of each bond (for instance, the first row should provide the flow of the amounts of -50 units of the first asset).*

Sol.: See Script `S_PortBonds`.

Exercise 18 (Portfolio of bonds 2) *Create a .xls file in which the first row represents the vector* $t = [0.5, 1, 1.5, 2, 2.5, 3, 3.5, 4, 4.5, 5]$ *(namely the dates of cash flows), while the other rows represent the cash flow of the five assets described in Exercise 17. Then, import the data contained in this .xls file in your Workspace.*

Sol.: See Script S_PortBonds_2.

Exercise 19 (Element-by-element operations) *Generate a vector b of dimension N with at least twenty elements (N ≥ 20). This vector must have the components such that* $b_i = (-1)^{i+1}$, *i.e.,* $b = (1, -1, 1, ..., (-1)^{N+1})$. *Then, modify the elements of this vector with the indices multiple of 3, so that* $b_{3i} = 0$ *with* $i = 1, ..., N/3$. *Finally, modify the vector b, so that the last component is equal to 100.*

Sol.: See Script S_ElByElOp.

Exercise 20 (El-by-El operations and predefined functions)
Calculate the absolute value of the elements of $x = [7, -7]$ *by using the function* sign *(you should not use the built-in function* abs*).*

Sol.: See Script S_Absolute_value.

Exercise 21 (Integer part of numbers) *Round the scalar* $x = 2.32$
(i) to the closest integer number (see round*); (ii) to the closest integer number that is smaller than x (see* floor*); to the closest integer number that is greater than x (see* ceil*).*

Sol.: See Script S_Integer_Part.

Exercise 22 (Manipulation of arrays) *Create:*

1. *a 4×4 diagonal matrix, where the principal diagonal elements are random numbers uniformly distributed in the range* $[0, 1]$ *(use* rand *and* diag*);*

2. *a matrix with the upper-diagonal of 5 random elements uniformly distributed in the range* $[0, 1]$;

3. *a matrix containing 3 copies of the matrix of point 1 in the row di-*

mension, and check its dimension (use `repmat` *and* `size`*);*

4. *a matrix containing 2 copies of the matrix of point 2 in the column dimension, and check its dimension;*

5. *a matrix with all elements equal to 0 having the same dimension of the matrix of point 4 (use* `zeros`*);*

6. *a $n \times n$ matrix G in which $G(1, n) = 800$. Do it for several values of n.*

Sol.: See Script S_Manipulation_Arrays.

Exercise 23 (Manipulation of arrays 2) *Given a "magic" matrix A of dimension 7×7 (use* `magic`*), create:*

1. *the vectors x and y that represent the principal diagonal and the counterdiagonal of A, respectively. For y, use the built-in function* `fliplr`*. Furthermore, check that the sum of x elements is equal to the sum of y elements (use* `sum`*);*

2. *a set of vectors v_k with $k = 0, 1, 2, 3, 4, 5, 6$ (each with dimension 7), where the vector k represents a diagonal of order k of A (for instance, $v_1 = a_{1,2}; a_{2,3}; a_{3,4}; a_{4,5}; a_{5,6}; a_{6,7}; a_{7,1}$). Furthermore, check that the sum of the elements of each vector v_k is the same.*

Sol.: See Script S_Manipulation_Arrays_2.

Exercise 24 (Various arrays) *Solve the following items:*

1. *create two row vectors of dimension 1×10, one with elements from 1 to 10, and the other one with elements from 5 to 50 (with a step equal to 5). Hint: see colon (:) operator;*

2. *using the vectors of point 1, create two 3×10 matrices by replicating the vectors three times. Then, create another two 10×7 matrices by transposing and replicating the vectors of point 1.*

3. *Verify the dimensions of the matrices of points 1 and 2.*

Sol.: See Script S_Various_arrays.

Exercise 25 (Various arrays 2) *Solve the following points:*

1. *create a 3×5 matrix A having uniformly distributed random numbers. Then, check the dimensions of A;*

2. *given a vector x with 5 elements equal to 1 (both if x is a column vector and a row vector), how is it possible to perform the product A · x? To avoid errors, you could opportunely use the build-in function* reshape;

3. *check the dimensions of the array created in point 2.*

Sol.: See Script S_Various_arrays_2.

Exercise 26 (Various arrays 3) *Create the following arrays:*

1. *two row vectors of dimension 1 × 10. The first has elements from 1 to 10, while the second has elements from 10 to 100 with a step of 10.*

2. *Using the vectors of point 1, create two 5 × 10 matrices by properly replicating the vectors. Then, create another two matrices 10 × 7 by transposing and replicating the above vectors.*

3. *Check the dimensions of the arrays created in point 2.*

Sol.: See Script S_Various_arrays_3.

Exercise 27 (Elements of a matrix) *Given the matrix*

$$A = \begin{bmatrix} 7 & 2 & 6 & 1 \\ 3 & 7 & 5 & 2 \\ 5 & 9 & 1 & 3 \\ 4 & 8 & 2 & 1 \end{bmatrix},$$

1. *check which element of A is selected by the statements* A(2,3) *and* A(5). *Think carefully about their outputs;*

2. *separately select, from A, the first row, the last two columns, and the diagonal;*

3. *replace the first row with a vector having all elements equal to 2;*

4. *replace the second and third column with uniformly distributed random numbers;*

5. *set the element $a_{3,3} = 5$.*

Sol.: See Script S_Element_matrix.

Exercise 28 (Construction of a matrix) *Create a matrix A (12×6) according to the following steps.*

1. *Construct a matrix C (6×6) where the rows from 1 to 3 have all elements equal to 1, and the rows from 4 to 6 have all elements equal to 3.*

2. *Replace the first 5 elements of the second column with the elements of the first upper-diagonal of the "magic" matrix having the appropriate dimension. Then, replace the sixth element of this column with 0 (see* magic *and* diag *in the Help).*

3. *Multiply the third column by -2.*

4. *Create the final matrix by copying the matrix obtained in point 3 twice.*

Sol.: See Script S_Matrix_Construction_2.

Exercise 29 (Scalar product) *Given two vectors u and v with n elements, the scalar product is defined as $u \cdot v = \sum_{i=1}^{n} u_i \cdot v_i$. How is the scalar product implemented in MATLAB? For instance, compute the scalar product between the vector of asset prices* P=[10;15;50;20;25] *and the corresponding shares vector* q=[1000;500;300;600;700]. *From the financial viewpoint, what does this result represent?*
Sol.: See Script S_Scalar_product.

Exercise 30 (Matrix calculus) *Solve the following points.*

1. *Given a row vector v and an integer n, generate the matrix* repmat(v,n,1) *by using the build-in function* ones *and the matrix product* *.*

2. *Given a column vector w and an integer n, generate the matrix that*

can be obtained by `repmat(w,1,n)` *using the build-in function* `ones` *and the matrix product* `*`.

Sol.: See Script `S_Matrix_calc`.

Exercise 31 (Portfolio of stocks) *Let us consider a $n \times m$ matrix R_{1w} of weekly returns, where n is the number of simulations and m the number of stocks, and a $1 \times m$ vector P_0 that represents their current prices. Let us denote the simulations of the assets' prices by an $n \times m$ matrix P_{1w}. Recall that $R_{\tau} = \frac{P_{\tau}}{P_0} - 1$ and, therefore, $P_{\tau} = P_0(1 + R_{\tau})$. Consider the following data*

$$R_{1w} = \begin{bmatrix} 0.03 & 0.05 & 0.04 & 0.02 & 0.05 \\ 0.05 & 0.06 & 0.05 & 0.03 & 0.08 \\ -0.04 & -0.02 & 0.01 & -0.01 & -0.10 \end{bmatrix}$$

and $P_0 = [10\ 20\ 15\ 20\ 25]$, and compute the price matrix P_{1w}.
Furthermore, let $q = [1000\ 800\ 1100\ 700\ 500]$ be the stock shares in a portfolio. Compute the $n \times 1$ portfolio value vector $V_{q,1w}$ and the corresponding return vector $R_{q,1w} = \frac{V_{q,1w}}{V_0} - 1$. Save the results in a .txt file.
Note that the returns matrix R_{1w} can be generated by a Monte Carlo simulation. For example, let us consider a return matrix `R_MC=randn(n,m)/10` *with $n = 100$ and $m = 5$, that represents the simulations of a multivariate normal variable. Setting P_0 and q as before, compute P^{MC}, V_q^{MC}, R_q^{MC}.*

Sol.: See Script `S_Portfolio_Stocks`.

Exercise 32 (Expected return and variance of a portfolio)
Consider a market with n assets where $\mu = [\mu_1, \mu_2, ..., \mu_n]$ is the vector of the assets expected returns and $\Sigma = \{\sigma_{ij}\}$ with $i,j = 1, ..., n$ is the covariance matrix of the asset returns. For instance, consider $n = 5$, $\mu = [0.05, 0.1, 0.15, 0.08, 0.11]$, and

$$\Sigma = \begin{bmatrix} 0.1 & 0 & -0.05 & 0.3 & -0.7 \\ 0 & 0.2 & 0.15 & -0.1 & 0 \\ -0.05 & 0.15 & 0.5 & 0.2 & -0.15 \\ 0.3 & -0.1 & 0.2 & 0.3 & 0.25 \\ -0.7 & 0 & -0.15 & 0.25 & 0.4 \end{bmatrix}.$$

Solve the following points.

1. *Generate the vector x representing the Equally-Weighted portfolio, where the allocation of capital in the portfolio to be invested in each asset i is equal, namely $x_i = 1/n$, $\forall i$.*

2. *Compute the expected return of the Equally-Weighted portfolio.*

3. *Compute the variance of the Equally-Weighted portfolio.*

4. *Save the results in a .txt file.*

Sol.: See Script S_Portfolio_exp_ret_variance.

Exercise 33 (Euclidean Norm) *Given a vector x with n elements, its Euclidean norm is defined as*

$$\|x\|_2 = \sqrt{\sum_{i=1}^{n} x_i^2} \,.$$

Create a Function that computes the Euclidean norm of any vector x.
Sol.: See Function F_Euclidean_norm.

Exercise 34 (for loop) *Create a Script that computes the sum of the first $n = 100$ natural numbers using a for loop, and compare the result with the following closed-form solution due to Gauss:*

$$\sum_{j=1}^{n} j = \frac{n(n+1)}{2}$$

Furthermore, generate a Function that is able to solve the same problem for any n.

Sol.: See Script S_for_loop and Function F_for_loop.

Exercise 35 (for loop 2) *Solve the following problems using both for loops and matrix operations (or/and appropriate MATLAB built-in functions). Then, compare the results.*

1. *Let $n = 100$, generate the vector $w = \left[1, \frac{1}{2}, \frac{1}{3}, ..., \frac{1}{n}\right]$.*

2. Compute the sum of the elements of w (sum).

3. Compute the product of the elements of w (prod).

4. Compute the cumulative sum of the elements of w (cumsum).

5. Compute the cumulative product of the elements of w (cumprod).

6. Compute $\sum_{i=1}^{n} \frac{1}{i(i+1)}$ (for n sufficiently large it tends to 1).

7. Compute the partial sums $S(q) = \sum_{i=1}^{q} \frac{1}{i(i+1)}$ for $q = 1, \ldots, n$, and store them as the elements of a vector.

Sol.: See Script S_for_loop_2.

Exercise 36 (while loop) *Let us continue from Exercise 34. Create a Function that sums the first n natural numbers using a* while *loop.*

Sol.: See Function F_while_loop.

Exercise 37 (Matrix Generation) *Considering $n = 50$, solve the following items using matrix operations.*

1. *Create a $n \times n$ matrix J which provides the vector $[1, 2, \ldots, n]$ in each line. Then, save this matrix in the* matrixJ.txt *file.*

2. *Given a number $v = 3$, generate a $n \times n$ matrix A whose elements follow the rule $a_{ij} = v^{j-1}$. Finally, save A in the* matrixA.txt *file. Hint: use the matrix J of the previous point.*

3. *Given a number $v = 3$, create a $n \times n$ matrix B whose elements follow the rule $b_{ij} = v^{j-i}$. Then, save B in the* matrixB.txt *file. Hint: as before, use the matrix J.*

4. *In another Script, solve the previous three points using* for *loops.*

Sol.: See Script S_MatrixGeneration and S_MatrixGen_For.

Exercise 38 (Graphs) *Solve the following points.*

1. *Calculate the function* $f(x) = \dfrac{1}{\sqrt{2\pi}} e^{-\frac{x^2}{2}}$, *with* $x \in [-5, 5]$ *and make the graph.*

2. *Given a strike price* $K = 100$, *calculate the function*
   $c(S) = \max\{S - K, 0\}$ *with* $S \in [80, 120]$, *and make the graph. Note that* $c(S)$ *represents the payoff function of a call option.*

3. *Given a strike price* $K = 100$, *calculate the function*
   $p(S) = \max\{K - S, 0\}$ *with* $S \in [80, 120]$, *and make the graph. Note that* $p(S)$ *represents the payoff function of a put option.*

4. *Calculate the function* $f(x) = (x^3 - 2x + 1)\ln(x + 2)$ *with* $x \in [0, 5]$, *and make the graph.*

5. *Calculate the functions* $f(x) = \dfrac{e^{-\frac{x^2-1}{x+3}}}{x^2 - \cos(x)}$ *and* $g(x) = \dfrac{e^{-\frac{x^2-1}{x+3}}}{x^2 + \sin(x)}$, *where* $x \in [0, 1]$, *and then make the graph.*

For each graph, consider an appropriate step to discretize the interval where the independent variable lies. Furthermore, save each graph as a .jpeg file.

Sol.: See Script S_Various_Graphics.

Exercise 39 (Graphs of spreads and combinations options)
Solve the following points.

1. *Consider two call options on the same underlying stock* S *with the same expiration date, and with strike price* $K_1 = 90$ *and* $K_2 = 110$, *respectively. Calculate the function* $BullSpread(S) = \max\{S - K_1, 0\} - \max\{S - K_2, 0\}$ *with* $S \in [60, 140]$, *and make its graph showing also the two separate call options payoffs. Note that the strategy described above, known as* Bull Spread, *can be obtained buying the first option and selling the second one.*

2. *Consider two put options on the same underlying stock* S *with the same expiration date, and with strike price* $K_1 = 90$ *and* $K_2 = 110$, *respectively. Calculate the function* $BearSpread(S) = \max\{K_1 - S, 0\} - \max\{K_2 - S, 0\}$ *with* $S \in [60, 140]$, *and make its graph showing also the two separate put options payoffs. Note that the strategy described*

46

above, *known as* Bear Spread, *can be built buying the first option and selling the second one.*

3. *Consider three call options on the same underlying stock S with the same expiration date, and with strike price $K_1 = 90$, $K_3 = 110$ and $K_2 = \frac{K_1 + K_3}{2}$, respectively. Calculate the function $ButterfSpread(S) = \max\{S - K_1, 0\} + \max\{S - K_3, 0\} - 2\max\{S - K_2, 0\}$ with $S \in [60, 140]$, and make its graph showing also the three separate call options payoffs. Note that the strategy described above, known as Butterfly Spread, can be obtained buying the first two options and selling two of the third option.*

4. *Consider a put and a call option on the same underlying stock S with the same expiration date, and with the same strike price $K = 100$, respectively. Calculate the function $Straddle(S) = \max\{S - K, 0\} + \max\{K - S, 0\}$ with $S \in [60, 140]$, and make its graph showing also the two separate options payoffs. Note that the strategy described above, known as Straddle, can be built buying both the options.*

5. *Consider a put and a call option on the same underlying stock S with the same expiration date, and with strike price $K_1 = 90$ and $K_2 = 110$, respectively. Calculate the function $Strangle(S) = \max\{S - K_2, 0\} + \max\{K_1 - S, 0\}$ with $S \in [60, 140]$, and make its graph showing also the two separate options payoffs. Note that the strategy described above, known as Strangle, can be obtained buying both the options.*

Save each graph as a .jpeg file, naming the y-axis as "Payoff" and the x-axis as "Price", and showing the corresponding strategy name in the title.

Sol.: See Script S_Option_Strategy_Graphics.

Exercise 40 (Graphs 2) *Solve the following points.*

1. *In the Script* S_Various_Graphics_2, *calculate the function*

$$f(x) = \begin{cases} e^x - 1 & if \quad x \le 0 \\ x^2 & if \quad 0 < x \le 2 \\ x^3 - 4 & if \quad x > 2 \end{cases}$$

where $x \in [-3, 5]$, and make the graph, saving it as Various_Graphics_2.jpg.

47

2. *Define the vector* ext=[x_min, x_max], *and an integer* $N \geq 2$ *that indicates the number points between* x_{min} *and* x_{max}, *where the function must be calculated. Therefore, write a MATLAB Function, named* F_Various_Graphics_2, *to calculate* $f(x)$ *between* x_{min} *and* x_{max}, *specified in the vector* ext.

Sol.: See Script S_Various_Graphics_2 and Function F_Various_Graphics_2.

Exercise 41 (Vector construction - for loop) *Setting* $n = 100$, *solve the following items.*

1. *Using a* for *loop, generate a vector* $v = [v_1, v_2, \ldots, v_k, \ldots, v_n]$ *where the generic k-th element* $v_k = \frac{2^k}{k!}$. *Then, save this vector in a .txt file, named* vectorv.txt.

2. *Make the graph of* v_k *as a function of k with* $k = 1, \ldots, n$, *and save it as* v_graph.tif.

3. *In the Script* S_ForVector_calc, *solve the two previous points without using a* for *loop.*

Sol.: See Script S_ForVector and S_ForVector_calc.

Exercise 42 (Series - for loop) *Setting* $n = 100$, *solve the following points.*

1. *Compute the sum of the finite series* $s(n) = \sum_{k=1}^{n} (-1)^{k-1} \left(\frac{1}{k} \right)$.

2. *Make the graph of* $s(n)$ *as a function of n. Note that, if n is sufficiently large, then* $s(n)$ *must tend to* $\ln(2)$. *Finally, save the graph as* sn_graph.tif.

Sol.: See Script S_LogSeries.

Exercise 43 (Series - while loop) *Setting* $n = 100$, *solve the following problems in the Script* S_series_calc.

1. Compute the sum of the finite series $s(n) = \sum\limits_{k=1}^{n} \dfrac{(-1)^{k-1}}{2k-1}$.

2. Plot $s(n)$ as a function of n. Note that, if n is sufficiently large, then $s(n)$ must tend to $\dfrac{\pi}{4}$. Finally, save the figure as S_series.tif.

3. Given that $\dfrac{\pi}{4} = \sum\limits_{k=1}^{\infty} \dfrac{(-1)^{k-1}}{2k-1}$, write a function (F_Series) which has an approximation tolerance L as input such that $\left| \dfrac{\pi}{4} - s(N) \right| < L$, and which provides the number of iterations N needed to achieve the required approximation tolerance and the value $S = s(N)$ as outputs.

4. Using the Function F_Series, write a Script to compute the number of iterations N and the the value of $S = s(N)$, given an approximation tolerance $L = 10^{-8}$.

Sol.: See Script S_series_calc and Function F_series.

1.6 Exercises on the basics of financial evaluation

Below we propose several exercises concerning Financial Mathematics, which cover some of the main topics of this subject. Even though the mathematical definitions of all the variables are provided, we do not dwell on their economic and financial meaning. Therefore, we suggest, e.g., Buchanan (2012); Castagnoli et al (2013); Castellani et al (2005a); Cipra (2010) for further clarification and understanding. However, the definitions contained in this section are thorough enough to allow the reader to solve these exercises. The last part of this section deals with Interest Rate Swaps and relative exercises.

Exercise 44 (Discount factor) Let $t = [0.5, 1, 1.5, 2, 2.5, 3, 3.5, 4, 4.5, 5]$ be a schedule expressed in years.

1. Assuming a flat (annual) interest rate curve (for instance, $\bar{i} = 2\%$), generate a vector $\bar{v}$ that defines the values of the discount factors corresponding to the times of the schedule as follows: $\bar{v}_k = v(0, t_k) = (1 + \bar{i})^{-t_k}$ with $k = 1, 2, ..., 10$.

2. Assuming a non-flat (annual) interest rate curve $\tilde{i}$, compute the price term structure $\tilde{v}_k = \tilde{v}(0, t_k) = (1 + \tilde{i}(0, t_k))^{-t_k}$ with $k = 1, 2, ..., 10$.

For example, consider

$$\tilde{i} = [0.008, 0.010, 0.015, 0.018, 0.025, 0.031, 0.037, 0.04, 0.045, 0.047].$$

Save $\bar{v}$ and $\tilde{v}$ in a .txt file.

Sol.: See Script S_Schedule_Prices.

Exercise 45 (Present Value of cash flows) *Let us consider the schedule t, and the matrix C of Ex. 17 which represents assets' cash flows. Then, solve the following points.*

1. *Load the .txt file of Ex. 44, and compute the present value of the cash flows defined in C both with the flat and the non-flat yield curves.*

2. *Compute the duration of each asset at $t_0 = 0$.*

3. *Compute the duration of the portfolio which contains the following number of shares in each of its assets $q = [50, 100, 70, 80, 30]$.*

Sol.: See Script S_Present_Value.

Exercise 46 (Evaluation of Cash flows) *Solve the following points.*

1. *Write the Function F_flow_value that computes, at a generic time t, the value of the cash flow $X = [x_1, x_2, ..., x_n]$ both by a flat $\bar{i}$ and a floating $\tilde{i}$ interest rate curve. Let $s = [s_1, s_2, ..., s_n]$ be the schedule of the cash flow F, with the value at time t being:*

$$V(t, X) = \sum_{k=1}^{n} x_k \left(1 + \tilde{i}_k\right)^{-(s_k - t)}$$

2. *In the Script S_f_flow_value, test the Function F_flow_value for the following data: $\bar{i} = 5\%$, $X = [5, 5, 105]$, and $s = [1, 2, 3]$ (thus, $t = [0, 1, 2, 3]$). Hint: the length of the output V should be the same of t.*

3. *In the same Script, repeat the evaluation of F with the floating interest rate structure $\tilde{i} = [0.037, 0.042, 0.051]$.*

4. *Furthermore, save the results of points 2 and 3 in the file* `cash_flow_value.txt`.

Sol.: See Function `F_flow_value` and Script `S_f_flow_value`.

Exercise 47 (Net Present Value and Duration) *Let us consider a schedule $t = [t_1, t_2, ..., t_n]$ and an $m \times n$ matrix C, where m is the number of available assets, and n represents the number of due dates, that coincides with the length of t. Let $\tilde{i} = [i_1, i_2, ..., i_n]$ be the non-flat (annual) interest rate curve. Write a function $\mathbf{F\_Flow\_indexes\_npv}$ that computes the Net Present Value (NPV) vector of the investments defined in C, and the duration vector D at t_0. Note that the length of these two vectors is equal to m. Recall that the NPV and the duration of an investment $x = [x_1, x_2, ..., x_n]$ are respectively:*

$$NPV(t_0, x) = \sum_{k=1}^{n} x_k (1 + i_k)^{-(t_k - t_0)}$$

$$D(t_0, x) = \frac{\sum_{k=1}^{n} (t_k - t_0) x_k (1 + i_k)^{-(t_k - t_0)}}{NPV(t_0, x)}$$

where x_k is the amount of money at time t_k.

Sol.: See Function `F_Flow_indexes_npv`.

Exercise 48 (Net Present Value and Duration 2)
Given $t = [1, 2, 3, 4, 5]$,

$$C = \begin{bmatrix} 7 & 7 & 7 & 7 & 107 \\ 6 & 6 & 6 & 106 & 0 \\ 0 & 0 & 100 & 0 & 0 \\ 0 & 0 & 0 & 0 & 100 \end{bmatrix},$$

and $\tilde{i} = [0.05, 0.046, 0.044, 0.049, 0.052]$, test the Function of Ex. 47, thus computing the NPV and the duration of the four investments described in C.

Sol.: See Script `S_flow_indexes_npv`

Exercise 49 (Postponed Annuity) *Determine the initial value $V_0(i, n)$ of a postponed annuity, supposing an (annual) interest rate $\bar{i} = 12\%$, an annual payment $R = 10$, and a length of the annuity of $n = 17$ years. Remember that, in the case of annual payments, $V_0 = R(v + v^2 + \ldots + v^n)$, where $v = \dfrac{1}{1 + \bar{i}}$ is the annual discount factor. Calculate V_0 using a* `for` *loop, and verify that it coincides with the result obtained by the close-form expression $V_0 = Rv\dfrac{1 - v^n}{1 - v}$. Finally, make the graph of $V_0(n)$ as function of the length of the annuity n (e.g., $n = 1, \ldots, 100$), and check that the function $V_0(n)$ has a horizontal asymptote, which corresponds to the present value of a perpetual annuity $\overline{V_0} = \dfrac{R}{i}$.*

Sol.: See Script S_Postponed.

Exercise 50 (Amortization schedule) *Create a matrix that represents an amortization schedule with postponed payments, where the columns represent the schedule, the total payment R_k, the principal portion C_k, the interest portion I_k, and the principal balance (residual debt) D_k respectively, with $k = 1, \ldots, n$. Let us consider a French amortization schedule on a period of 10 years with semi-annual payments ($n = 20$), the debt $S = 140000$, and the (annual) interest rate $\bar{i} = 5.7\%$. We recall here some formulas and notations. For the French amortization, the total payments are constant, namely $R_k = R = \dfrac{S}{a_{n\neg\bar{i}}}$ for all k, where $a_{n\neg\bar{i}} = v\dfrac{1 - v^n}{1 - v}$, and $v = \dfrac{1}{1 + \bar{i}}$. Then, the output will be a 21×5 matrix with the following elements:*

Time	Total	Principal	Interest	Balance
0	0	0	0	$D_0 = S$
1	$R_1 = R$	$C_1 = R - I_1$	$I_1 = iS$	$D_1 = S - C_1$
$\vdots$	$\vdots$	$\vdots$	$\vdots$	$\vdots$
n	$R_n = R$	$C_n = R - I_n$	$I_n = iD_{n-1}$	$D_n = D_{n-1} - C_n$

Finally, save the output in the file `french.xls`.

Sol.: See Script S_Amortization_Schedule.

Exercise 51 (Amortization schedule 2) *Let us continue from Exercise 50. Compute a matrix that represents the output of an Italian amortization schedule (equal principal portions). Note that, in this case, $C_k = C = \dfrac{S}{n}$ and $D_k = S - \sum_{h=1}^{k} C_h$. Save the output in the file* `italian.xls`.

Sol.: See Script S_Amortization_Schedule_2.

Exercise 52 (Amortization schedule 3)
Appropriately combine the codes used in the Scripts of Exs. 50 and 51 in the Function F_Amortization_Schedule by means of an if *statement, in order to choose either an Italian or a French amortization schedule. For instance, you could insert flags equalling 0 and 1 for the French and Italian amortization schedules respectively. Finally, write a Script to apply* F_Amortization_Schedule *to both amortization schedules.*

Sol.: See Function F_Amortization_Schedule
and Script S_Amortization_Schedule_3.

Exercise 53 (Cash flows 2) *Write the Function F_flow_value_bis by which one can calculate the value of a cash flow $F = [f_1, f_2, ..., f_n]$, at time $t = 0$. Let us consider the schedule $s = [s_1, s_2, ..., s_n]$, and the yield to maturity specified by the following expression*

$$h(0, s_k) = \alpha + \beta s_k + \gamma s_k^2,$$

where $\alpha = 0.0024$, $\beta = 0.0097$, and $\gamma = 0.0033$. Furthermore, recall that

$$i(0, s_k) = \exp(h(0, s_k)) - 1 \quad and \quad V(0, F) = \sum_{k=1}^{n} f_k (1 + i(0, s_k))^{-s_k}.$$

Using the Function F_flow_value_bis in the Script S_flow_value_bis, compute the present value of the flow $F = [6, 6, 6, 6, 106]$, where the schedule $s = [1, 2, 3, 4, 5]$. Finally, save the result in a .xls file.

Sol.: See Function F_flow_value_bis and Script S_flow_value_bis.

Exercise 54 (Zero Coupon Bond) *Solve the following points.*

1. *Write the Function* F_zcb_return *that calculates the annual return i of a Zero Coupon Bond (ZCB) with face value 100. Let P be its issue price, and T the time to maturity (expressed in years), then*

$$i = \left(\frac{100}{P}\right)^{\frac{1}{T}} - 1.$$

2. *Consider the vector P = [99.88, 99.85, 99.76, 99.24, 97.33], representing the prices of 3-month, 4-month, 6-month, 1-year and 2-year ZCBs. Thus using* F_zcb_return, *compute the annual return of these ZCBs in the Script* S_zcb_return.

3. *Save the vector of the annual returns i in the file* Annual_return.txt.

Sol.: See Function F_zcb_return and Script S_zcb_return.

Exercise 55 (Forward price and forward rate) *Write a Function named* F_forward.m, *where the inputs are the schedule* $t = [t_1, t_2, ..., t_n]$ *and the vector of unit ZCB prices* $v = [v(0, t_1), v(0, t_2), ..., v(0, t_n)]$. *Note that the unit ZCB prices represent the term structure of prices. The outputs of the Function are the forward prices and the (annual) forward rates. Recall that the expression for forward prices is*

$$v(0, t_{k-1}, t_k) = \frac{v(0, t_k)}{v(0, t_{k-1})} \quad \text{with} \quad k = 2, 3, ..., n,$$

where n is the length of vector v, and $v(0, 0, t_1) = v(0, t_1)$. *Thus, the inputs and the outputs must have the same dimension. Furthermore, recall that the expression of the (annual) forward rates is*

$$i(0, t_{k-1}, t_k) = \left(\frac{1}{v(0, t_{k-1}, t_k)}\right)^{\frac{1}{t_k - t_{k-1}}} - 1 \quad \text{with} \quad k = 2, 3, ..., n,$$

where $i(0, 0, t_1) = \left(\frac{1}{v(0, 0, t_1)}\right)^{\frac{1}{t_1}} - 1.$

Sol.: See Function F_forward.

Exercise 56 (Forward price and forward rate 2) *Given a schedule* $t = [0.5, 1, 1.5, 2]$ *and a vector of ZCB prices* $v = [0.98, 0.96, 0.94, 0.93]$, *solve the following points.*

1. *Compute the (annual) spot interest rates. Recall that the (annual) spot rates are described by the expression* $i(0, t_k) = \left(\dfrac{1}{v(0, t_k)}\right)^{\frac{1}{t_k}} - 1$ *with* $k = 1, \ldots, n$.

2. *Using the Function* F_forward.m *of Ex. 55, calculate the forward rates.*

3. *In the same graph, plot spot and forward rates. Then, save the graph as* rates_graph.tif.

Sol.: See Script S_fwd_pricerate.

Exercise 57 (Internal Rate of Return IRR) *Consider a financial operation described by the vector of cash flows* $x = [-100, 2.5, 2.5, 102.5]$ *and by the vector of times* $t = [0, 6, 12, 18]$, *where the elements of vector t are expressed in months. In the Script* S_IRR, *compute the (annual) Internal Rate of Return (IRR) of this financial operation, and express it in percentage. More specifically, compute IRR by finding the zeros of the following equation:*

$$g(v) = x_0 + x_1 v^1 + \ldots + x_m v^m = \sum_{k=0}^{m} x_k v^k = 0. \qquad (1.2)$$

Furthermore, define a function $f(v) = \sum_{k=1}^{m} x_k v^k$ *and a straight line* $y = P = -x_0$, *and plot them in the same graph. What does the intersection between* $f(v)$ *and* y *represent? Hint: to find the zeros of* $g(v)$, *use the built-in function* fzero.

Sol.: See Script S_IRR.

1.6.1 Interest Rate Swap

An Interest Rate Swap (IRS) is a derivative contract in which two entities agree to perform an exchange of interest rates referred to the same notional capital. The exchange of interest rates is from a fixed interest rate to a floating interest rate (or vice versa). For instance, let us assume that we pay floating interest rates and receive fixed interest rates as in Fig. 1.3. Then, we can express the

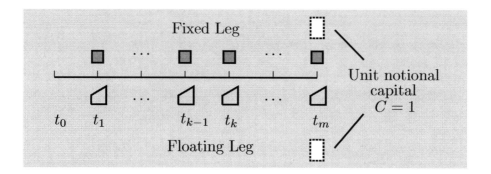

Figure 1.3: Scheme of a Vanilla Interest Rate Swap.

value of IRS at the starting time t_0 as the difference between the value of the IRS fixed leg (*FixL*) and the value of its floating leg (*FloL*):

$$V(t_0, IRS) = V(t_0, FixL - FloL) = \underbrace{V(t_0, FixL)}_{\text{fixed leg}} - \underbrace{V(t_0, FloL)}_{\text{floating leg}}.$$

According to the *no-arbitrage* condition, $V(t_0, IRS) = 0$, thus $V(t_0, FixL) = V(t_0, FloL)$. Without loss of generality, let us consider a unit notional capital, which is not exchanged. This implies that $V(t_0, FixL) = z_m v(t_0, t_1) + z_m v(t_0, t_2) + \ldots + z_m v(t_0, t_m) + v(t_0, t_m)$, while since the floating leg cash flow can be considered as the cash flow of a coupon bond with floating rates that is valued at par, $V(t_0, FloL) = 1$. Therefore, we obtain that

$$z_m \sum_{k=1}^{m} v(t_0, t_k) = 1 - v(t_0, t_m)$$

$$z_m = \frac{1 - v(t_0, t_m)}{\sum_{k=1}^{m} v(t_0, t_k)}. \tag{1.3}$$

Let us express Equation (1.3) for different m, namely

$$
\begin{array}{llll}
\text{for} & m = 1 & (1 + z_1)v(t_0, t_1) & = 1 \\
\text{for} & m = 2 & z_2 v(t_0, t_1) + (1 + z_2)v(t_0, t_2) & = 1 \\
\text{for} & m = 3 & z_3 v(t_0, t_1) + z_3 v(t_0, t_2) + (1 + z_3)v(t_0, t_3) & = 1 \\
& \vdots & \vdots & \vdots
\end{array} \tag{1.4}
$$

56

Thus, in matricial form we have

$$
\begin{pmatrix}
1+z_1 & 0 & 0 & \cdots & 0 & 0 \\
z_2 & 1+z_2 & 0 & \cdots & 0 & 0 \\
z_3 & z_3 & 1+z_3 & \cdots & 0 & 0 \\
\vdots & \vdots & \vdots & \ddots & \vdots & \vdots \\
z_{m-1} & z_{m-1} & z_{m-1} & \cdots & 1+z_{m-1} & 0 \\
z_m & z_m & z_m & \cdots & z_m & 1+z_m
\end{pmatrix}
\begin{pmatrix}
v(t_0,t_1) \\
v(t_0,t_2) \\
v(t_0,t_3) \\
\vdots \\
v(t_0,t_{m-1}) \\
v(t_0,t_m)
\end{pmatrix}
=
\begin{pmatrix}
1 \\
1 \\
\vdots \\
\vdots \\
1
\end{pmatrix}
$$
(1.5)

Let us denote the swap rates matrix with A, the vector of prices with v, and the vector of ones with $\mathbf{1}$. Then, the system of linear equations (1.5) can be synthetically written as:

$$
Av = \mathbf{1} \quad \Rightarrow \quad v = A^{-1}\mathbf{1}
$$
(1.6)

due to the non-singularity of A. Indeed, $\det A = \prod_{k=1}^{m}(1+z_k) \neq 0$.

Exercise 58 (Bootstrapping) *Consider a generic vector $z = [z_1, z_2, ..., z_n]$ and the matrix*

$$
A =
\begin{bmatrix}
1+z_1 & 0 & 0 & \cdots & 0 & 0 \\
z_2 & 1+z_2 & 0 & \cdots & 0 & 0 \\
z_3 & z_3 & 1+z_3 & \cdots & 0 & 0 \\
\vdots & \vdots & \vdots & \ddots & \vdots & \vdots \\
z_{m-1} & z_{m-1} & z_{m-1} & \cdots & 1+z_{m-1} & 0 \\
z_m & z_m & z_m & \cdots & z_m & 1+z_m
\end{bmatrix}
$$

as in (1.5) and (1.6). Let $z = [0.021, 0.025, 0.032, 0.043]$, then solve the following points.

1. *Create the matrix A. Hint: use the build-in MATLAB functions* `repmat`, `eye`, `tril`.

2. *Applying Expression (1.6), compute the term structure of prices v (bootstrapping), and verify that $Av = [1, 1, ..., 1]'$.*

Sol.: See Script S_Bootstrap.

Exercise 59 (Swap rates) *Let us consider a swap rate vector*
$z = [z_1, z_2, ..., z_m]$, *a vector* $v = [v_1, v_2, ..., v_m]$ *that represents the term structure of ZCB prices, and a time vector* $t = [1, 2, ..., m]$. *Solve the following problems.*

1. *Using the procedure described in Ex. 58, create a Function* F_Bootstrap *to obtain the vector* v, *given the vector* z.

2. *Supposing that* $z = [0.0185, 0.0223, 0.0297, 0.0313]$, *compute the vector* v *with the Function* F_Bootstrap *and with a* for *loop, using the iterative procedure, called bootstrapping, via the well-known relationship between swap rates and ZCB spot prices*

$$v_k = \frac{1 - z_k \sum_{j=1}^{k-1} v_j}{1 + z_k} \quad with \quad k = 1, \ldots, m.$$

3. *Using an arbitrarily chosen price vector* v, *calculate the swap rate vector* z *with a* for *loop considering the following relationships*

$$z_k = \frac{1 - v_k}{\sum_{j=1}^{k} v_j} \quad with \quad k = 1, \ldots, m.$$

Sol.: See Function F_Bootstrap and Script S_Swap_Rates.

Exercise 60 (IRS Plain Vanilla) *In the file* IRS_plain_vanilla.xls, *one can find the data of Interest Rate Swap (IRS) plain vanilla[a] corresponding to different maturities, namely the vector of various times to maturity* $t = [1, ..., T]$ *and the vector of swap rates* $z = [z_1, z_2, ..., z_T]$. *In the Script* S_IRS_PV, *load* t *and* z; *then, compute and compare spot and forward interest rates, recalling the Functions required to solve the following points.*

1. *Given the vector* z, *compute the vector* v, *representing the term structure of spot prices, with the Function* F_Bootstrapp *of Ex. 59.*

2. *Write the function* F_StructInterYield *to calculate the vector* y *of the yield to maturities*

$$y(i) = -\ln(v(i))/t(i),$$

with $i = 1, \ldots, T$. *Then, make a linear interpolation of these values to obtain the vector y_{int} of the yield to maturities corresponding to the times $t_{int} = [0.5, 1, 1.5, \ldots, T]$. Finally, using the interpolated values y_{int}, compute the spot prices vector v_{int} by the following formula*

$$v_{int}(i) = \exp\left(-y_{int}(i) * t_{int}(i)\right).$$

Hint: see the built-in function `interp1` *in the MATLAB Help.*

3. *Using the newly created* **F_StructInterYield** *and the Function* **F_forward** *of Ex. 55, compute the forward prices and the forward rates corresponding to v_{int}.*

4. *Make the graph of spot and forward rates as a function of time using different colors, and insert a legend to specify them.*

5. *Repeat Points 3, 4 directly interpolating the swap rates curve (corresponding to the time vector t_{int}), identified by the vector z.*

Sol.: See Script **S_IRS_PV** and Functions **F_StructInterYield**, **F_forward**.

---

[a]It is an IRS with a float and a fixed leg on the same notional principal amount, with a single currency and fixed payment dates.

Part II

Portfolio selection

Chapter 2

Preliminary elements in Probability Theory and Statistics

Since most financial models represent random phenomena, in this chapter we give a short introduction to the basic elements of Probability Theory and Statistics, which are the typical mathematical tools used in this framework. More precisely, in Section 2.1 we provide some preliminary elements in probability. Section 2.2 introduces the concept of discrete random variables, while Section 2.3 shows how to characterize a random variable by means of its cumulative distribution function. In Section 2.4 we introduce continuous random variables, describing how to characterize them through their probability density functions. In Section 2.5 we discuss some standard synthetic indices of a distribution. Finally, Section 2.6 is devoted to present some of the most frequently used probability distributions in finance.

2.1 Basic concepts in probability

Firstly, we seek to define the probability associated to a specific event. In the *Classical* (Bernoulli and Laplace) definition, the probability of an event is the ratio between the number of cases favorable to it and the number of possible cases, which should be considered equally likely. In the *Frequentist* definition, the probability of an event is the limit of the frequency of the occurrence of such an event when the number of trials is large (i.e., it tends to infinity). In the *Subjective* definition (Ramsey, De Finetti, Savage), the probability can be considered as the degree of confidence that an event will occur. It is determined by the information available to the individual and by his experience.

Remark 61 (Different approaches to define probability) *The* Classical *definition can be applied only when all the events have the same probability of occurrence (a situation that is generally not the case in financial contexts). The* Frequentist *definition can be used when the experiment is repeated multiple times under the same exact conditions (this is seldom the case in finance). In addition, it is not clear how to treat the concept of the limit to infinity of the frequency of the occurrence of an event. Both of these limitations can be overcome by the* Subjective *approach, but nonetheless with this definition it is impossible to assign a unique probability to an event, since different individuals can provide different estimations of the probability.*

Remark 62 *In view of the nature of the events in finance, we use the* Subjective *approach. However, independently of the approaches used to determine the probabilities of the elementary events, such probabilities have to satisfy logical requests of coherence, the so-called Kolmogorov's axioms. From these axioms, it is possible to obtain further properties, rules and theorems for computing probabilities (see, e.g., Gnedenko, 2018).*

Example 63 (De Finetti) *Consider a football match with the following possible events:* $1, \times, 2$ *(1 for the home win, $\times$ for the draw, and 2 for the away win). The probability of the home team's victory becomes:*

- *in the* Classical *approach:* $p = 1/3$

- *in the* Frequentist *approach, a large number of historical events are analyzed, and then* $p = \dfrac{n° \, of \, wins}{n° \, of \, matches \, played}$

- *in the* Subjective *approach, the physical conditions of the players are taken into account, as well as the state of the field, and so on. Then, the subjective probabilities are determined.*

To introduce some basic concepts of Probability Theory, as usual we consider the realizations of the throwing dice as an example of discrete events.

As shown in the following example, when the probability of each elementary event is assigned, one could ask to determine the probability related to a more complex event with respect to the elementary ones.

Example	Description	Concept
$\{1, 2, 3, 4, 5, 6\}$	space of the events	$\Omega = \{\omega_1, \omega_2, \ldots, \omega_n\}$
e.g., 4	elementary event	ω_i
$p(1), p(2), \ldots, p(6)$	prob of the event ω_i occurring	$p(\omega_i) = p_i$

Table 2.1: Example of discrete probability space

Example 64 (throwing dice) *Consider the example of discrete random events summarized in Table 2.1, and determine the probability of rolling an even number.*

Note that the event "even number" can be mathematically represented as a subset E of the space of the events $\Omega = \{1, 2, 3, 4, 5, 6\}$:

$$E = \{2, 4, 6\} \subset \Omega .$$

Then, the probability of rolling an even number is:

$$p(E) = \frac{1}{6} + \frac{1}{6} + \frac{1}{6} = \frac{1}{2} .$$

Furthermore, determine the probability of rolling a number less than 3. Let us define the event "number less than 3 " by $F = \{1, 2\} \subset \Omega$. Then, the probability of rolling a number less than 3 is:

$$p(F) = \frac{1}{6} + \frac{1}{6} = \frac{1}{3} .$$

More generally, for a random event represented by a subset of elementary events E belonging to the *sample* space Ω we have the following definition.

Definition 65 *Let $E \subseteq \Omega$. We say that E is an event and $P(E)$ is its probability measure (determined by means of the elementary event probabilities $P(\omega_i)$). By convention $0 \leq P(E) \leq 1$. Specifically, the probability measure P on E is an application $P : E \to [0, 1]$. Then $P(E) = mis(E)$, in particular $P(\Omega) = mis(\Omega) = 1$.*

Therefore, the events will be characterized as all the possible subset of Ω along with the empty set $\emptyset$ and the sample set Ω with $P(\emptyset) = 0$ and $P(\Omega) = 1$. In other words, the events can be represented by subsets of Ω, and, thus it makes sense to use set operations on events (e.g., union, intersection, complement) to define further new events.

Example 66 *Let us continue from Example 64. The events of the throwing of a regular six-sided dice are:*

$$\{\} = \emptyset$$
$$\{1\}, \{2\}, \dots, \{6\}$$
$$\{1,2\}, \{1,3\}, \dots, \{5,6\}$$
$$\{1,2,3\}, \dots, \{4,5,6\}$$
$$\{1,2,3,4\}, \dots, \{3,4,5,6\}$$
$$\{1,2,3,4,5\}, \dots, \{2,3,4,5,6\}$$
$$\{1,2,3,4,5,6\} = \Omega,$$

that are all possible subsets of $\Omega = \{1,2,3,4,5,6\}$, *i.e., the power set of* Ω. *It is possible to show that the power set contains* $2^{|\Omega|} = 2^6$ *subsets, including the empty set and* Ω *itself. We denote by* $|\Omega|$ *the cardinality of* Ω, *that, in a nutshell, is the number of the elementary events of* Ω.
Now, consider the following events

$$E_1 = \{1,2,3\} \quad and \quad E_2 = \{2,4,6\}.$$

Note that doing operations on sets (e.g., union and intersection) makes sense since the events are sets. Furthermore, it could be interesting to understand the meaning of their probability measure. For instance, let us consider the event "occur E_1 *or* E_2", *i.e.,* $E_1 \cup E_2 = \{1,2,3,4,6\}$, *and the event "occur* E_1 *and* E_2", *i.e.,* $E_1 \cap E_2 = \{2\}$. *For these new events, we have that* $P(E_1 \cup E_2) = \dfrac{5}{6}$ *and* $P(E_1 \cap E_2) = \dfrac{1}{6}$.

In the following remark we briefly introduce the mathematical concept of a probability space that models random events. It consists of three ingredients: a set Ω of all possible outcomes of an experiment; a set of events $\mathcal{F}$, where each event is a subset of Ω; a probability function P that attributes a *probability* to each event belonging to $\mathcal{F}$.

Remark 67 (Probability space) *In order to define a probability space, we could start by indicating the initial set of elementary events* Ω, *namely the space of events (see Examples 64 and 66). Then, we should consider a family* $\mathcal{F}$ *of subsets of* Ω, *where* $\mathcal{F}$ *has to satisfy appropriate conditions. The elements of the family* $\mathcal{F}$ *represent random events and* $\mathcal{F}$ *is called* σ*-algebra of events. The last ingredient is represented by the probability measure* P *defined on each element of* $\mathcal{F}$. *Formally, we have that* $P : E \in \mathcal{F} \to [0,1]$. *Together these three parts* $\{\Omega, \mathcal{F}, P\}$ *are called probability space.*

Now, we introduce some further basic concepts of probability, in which, as mentioned above, random events are represented by sets. Therefore, new events (i.e., sets) can be obtained by using basic operations of set theory (union, intersection, etc.).

Theorem 68 (probability of the sum of two events) *Given two events E_1 and E_2, we have $P(E_1 \cup E_2) = P(E_1) + P(E_2) - P(E_1 \cap E_2)$.*

Proof. For an intuitive proof see Fig. 2.1. ∎

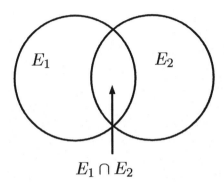

$$E_1 \cap E_2$$

Figure 2.1: Example of compatible events

Example 69 *Let us continue from Example 66. Given $E_1 = \{1, 2, 3\}$ and $E_2 = \{2, 4, 6\}$, we have $E_1 \cup E_2 = \{1, 2, 3, 4, 6\}$ and $E_1 \cap E_2 = \{2\}$, then*

$$\begin{aligned} P(E_1 \cup E_2) &= P(E_1) + P(E_2) - P(E_1 \cap E_2) \\ &= \frac{1}{2} + \frac{1}{2} - \frac{1}{6} = \frac{5}{6}. \end{aligned}$$

Definition 70 (incompatible events) *Given two events (sets) E and F, if $E \cap F = \emptyset$, then E and F are said incompatible.*

Remark 71 *If E and F are incompatible, $P(E \cup F) = P(E) + P(F)$. See Fig. 2.2.*

Example 72 *Let us continue from Example 69. It is clear that the event $E_2 = \{2, 4, 6\}$ could be seen as $E_2 = \{2\} \cup \{4\} \cup \{6\}$, in which case $P(E_2) = P(2) + P(4) + P(6) = \frac{1}{2}.$*

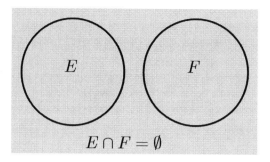

$$E \cap F = \emptyset$$

Figure 2.2: Example of incompatible events

Remark 73 (Conditional probability) *Let us introduce the concept of conditional probability (particularly important in finance) by means of an example. Given two events $F = \{2, 4, 6\}$ and $E = \{2\}$, as shown previously, $P(F) = \frac{1}{2}$ and $P(E) = \frac{1}{6}$. However, if we know that the event F has already occurred, the probability that the event E occurs, is influenced by this information. Let $P(E|F)$ denote the probability of the event E given the occurrence of the event F. In the above example one can easily calculate $P(E|F)$ using the classical definition of probability as the ratio between the number of favorable cases and the number of possible cases, namely $P(E|F) = \frac{1}{3}$.*

In more general terms, we can calculate the conditional probability using the following definition due to Kolmogorov.

Definition 74 (Kolmogorov's definition) *Let $E, F \subset \Omega$ two events defined on the same probability space $\{\Omega, \mathcal{F}, P\}$, the conditional probability of E given F is defined as the ratio of the probability of the joint events, E and F, and the probability of F:*

$$P(E|F) = \frac{P(E \cap F)}{P(F)} \ . \tag{2.1}$$

This definition can be easily interpreted using the classical definition of probability

$$
\begin{aligned}
P(E|F) &= \frac{N_{favorable}(E \cap F)}{N_{possible}(F)} = \\
&= \frac{N_{favorable}(E \cap F)}{N_{possible}(\Omega)} \frac{N_{possible}(\Omega)}{N_{possible}(F)} \\
&= \frac{P(E \cap F)}{P(F)}.
\end{aligned}
$$

Therefore, $P(E|F)$ represents an adjustment to the expectations of E due to the occurrence of F.

Example 75 *Let us continue from Remark 73.*
We have that $\Omega = \{1, 2, 3, 4, 5, 6\}$, $E = \{2\}$, $F = \{2, 4, 6\}$, *and* $E \cap F = \{2\}$.
Accordingly, $P(E \cap F) = \dfrac{N_{favorable}(E \cap F)}{N_{possible}(\Omega)} = \dfrac{1}{6}$, $P(F) = \dfrac{N_{favorable}(F)}{N_{possible}(\Omega)} = \dfrac{3}{6}$, *this means that following Relation* (2.1) *we can write that* $P(E|F) = \dfrac{P(E \cap F)}{P(F)} = \dfrac{1/6}{1/2} = \dfrac{1}{3}$, *as obtained in Remark 73.*

From the definition of conditional probability, we can intuitively interpret the concept of independence among events.

Definition 76 (independent events) *Two events E and F are said to be statistically independent if and only if $P(E \cap F) = P(E)P(F)$.*

This means that two events E and F are independent if and only if the probability of the joint events E and F is equal to the product of the probability of E and that of F. The following theorem shows how the definition of independence between events takes place when the occurrence of an event does not involve the probability of occurrence of the other.

Theorem 77 *If E and F are independent events, then $P(E|F) = P(E)$ or, equivalently, $P(F|E) = P(F)$.*

Proof. This intuitive result can be easily demonstrated using Definitions 74 and 76.

$$
\begin{aligned}
P(E|F) &= \frac{P(E \cap F)}{P(F)} \\
&= \frac{P(E)P(F)}{P(F)} = P(E).
\end{aligned}
$$

Similarly we can obtain that $P(F|E) = P(F)$. ∎

In general, $P(E|F) \neq P(F|E)$ and their relationship is expressed by Bayes' theorem.

Theorem 78 (Bayes' theorem) *Let E and F two random events, belonging to the same probability space $\{\Omega, \mathcal{F}, P\}$, then*

$$
P(E|F) = \frac{P(F|E)P(E)}{P(F)} . \tag{2.2}
$$

Proof. From Definition 74 of conditional probability we have

$$P(E|F) = \frac{P(E \cap F)}{P(F)} \ . \tag{2.3}$$

Analogously, $P(F|E) = \dfrac{P(F \cap E)}{P(E)}$. Since $P(E \cap F) = P(F \cap E)$, then

$$P(E \cap F) = P(F|E)P(E) \ . \tag{2.4}$$

Substituting (2.4) in (2.3), we obtain Expression (2.2). ∎

Remark 79 (Bayesian approach to estimation) *Theorem 78 is the core of the Bayesian method for estimating model parameters, an approach widely used in finance (see, e.g., Rachev et al, 2008, and references therein). Although a thorough discussion of the subject is beyond the scope of this book, we give here some insights that are behind this approach using the Bayes' rule (2.2). Suppose that E represents an evidence coming from the experience of an investor (prior), and that F represents an evidence obtained from the market data. Relation (2.2) states that, after the examination of the market data F, the investor's view E is adjusted according to (2.2). P(E) is named the prior probability, while the adjusted probability P(E|F) is named the posterior probability. In a nutshell, the posterior probability embodies both the investors view (prior), and the information from the market.*

Remark 80 (Space of the states of nature) *To introduce some preliminary concepts on Probability, in Example 64 we have shown a case where the space of events Ω consists of a finite number of elements (i.e., Ω = {1, 2, 3, 4, 5, 6}). As described above, any subset E ⊆ Ω represents a random event, and P(E) defines its probability of occurrence. However, this definition of events can lead to some issues when the space of the states is infinite, and thus a more precise definition of an event is necessary. Indeed, for this purpose only measurable subsets of Ω, constituting a σ-algebra over the space of events Ω, are considered events. For more details see Gnedenko (2018).*

Below, we provide some examples where the space of elementary events can have an infinite number of elements, or non-numeric elementary events.

Example 81 (Space of events with infinite elements)
Consider as random events the number of cars that will refill fuel in a specific station, and denote by ω the generic elementary event. The space of events

---

Ω *will be composed by*

$$\left.\begin{array}{l} \omega_0 \to \text{0 cars} \\ \omega_1 \to \text{1 car} \\ \omega_2 \to \text{2 cars} \\ \vdots \\ \omega_n \to \text{n cars} \\ \vdots \end{array}\right\} \Rightarrow \Omega = \{\omega_0, \omega_1, \omega_2, \ldots, \omega_n, \ldots\} = \{0, 1, 2, \ldots, n, \ldots\}.$$

As further example, let us consider as an elementary event ω the daily income of the above fuel station. In this case,

$$0 \le \omega < +\infty \Rightarrow \Omega = [0, +\infty) = \mathbb{R}^+.$$

Example 82 (Non-numeric space of the events) *Finally, we could consider the daily weather conditions as elementary events. For instance,*

$$\left.\begin{array}{rcl} \omega_1 & = & rain \\ \omega_2 & = & sun \end{array}\right\} \Rightarrow \Omega = \{\omega_1, \omega_2\}.$$

Unlike the above examples, here the space of events Ω consists of non-numeric random states of nature.

2.2 Random variables

In this section we introduce the concept of random variable. To simplify the discussion, as typically done overall this book, we start from a simple practical example.

Let X be the price that an equity will assume tomorrow, knowing that today its value is 1€. Supposing that the future price could generate an increase (↑) or a decrease (↓) of the equity value, tomorrow the states of nature can be $\omega_1 = \omega_\uparrow$ and $\omega_2 = \omega_\downarrow$. Therefore, we can define the space of events $\Omega = \{\omega_1, \omega_2\}$. Thus, the value of X is a function of the state that could occur tomorrow:

$$X(\omega) = \left\{ \begin{array}{rcl} X(\omega_\uparrow) & = & 2.00 \, \text{€} \\ X(\omega_\downarrow) & = & 0.50 \, \text{€} \end{array} \right.$$

Suppose that the probabilities associated with the elementary events are $P(\omega_\uparrow) = 1/3$ and $P(\omega_\downarrow) = 2/3$. We can extend the concept of the probability of an event to the probability that a random variable assumes a certain value as follows

$$P(\omega_\uparrow) = P(X = 2) = 1/3$$
$$P(\omega_\downarrow) = P(X = 0.5) = 2/3$$
$$P(X \neq 0.5, 2) = 0.$$

Thus, we could qualitatively say that X is a random variable, namely a quantity whose values depend on the case and for which the probabilities of occurrence are defined. More generally, let $\Omega = \{\omega_1, \omega_2, \ldots, \omega_n\}$ denote the space of discrete events and $p(\omega_i) = p_i$ (with $i = 1, \ldots, n$) the probabilities of the elementary events.

Definition 83 (discrete random variable) *Let $X : \Omega \to \mathbb{R}$ be a function of $\omega \in \Omega$ that assumes discrete values $X(\omega_1), X(\omega_2), \ldots, X(\omega_n)$ with probability $p_1, p_2, \ldots, p_n$. Then, X is called discrete random variable.*

Let us consider the r.v. X described at the beginning of this section. We can define the expected value of the equity price X as $E[X] = p_\uparrow X(\omega_\uparrow) + p_\downarrow X(\omega_\downarrow)$, namely the weighted sum of the values of X with the respective probabilities of realization. Note that $E[X]$ could be known even if we do not know which future event ω will be realized. Furthermore, since $E[X] = 1 \neq 0.5, 2$, the expected value can be different from any realization of X.

Definition 84 (Expected value) *Let X be a discrete random variable. The expected value of X is $E[X] = \sum_{i=1}^{n} p(\omega_i) X(\omega_i)$. If we denote $p(\omega_i) = p_i$ and $X(\omega_i) = x_i$, then $E[X] = \sum_{i=1}^{n} p_i x_i$.*

Remark 85 (Properties of the expected value) *Let X and Y be two random variables defined on the same probability space $\{\Omega, \mathcal{F}, P\}$, $X, Y : \Omega \to \mathbb{R}$. It is straightforward to show that the expected value is a linear operator:*

$$linearity = \begin{cases} additivity \quad \to \quad E[X + Y] = E[X] + E[Y] \\\\ homogeneity \quad \to \quad E[\alpha X] = \alpha E[X] \quad \forall \alpha \in \mathbb{R} \end{cases}$$

In particular, for any affine transformation $Z = aX + b$ with $a, b \in \mathbb{R}$, we have

$$E[Z] = E[aX + b] = aE[X] + b.$$

For instance, in the case of discrete random variables we can write

$$
\begin{aligned}
E[Z] &= E[aX + b] \\
&= \sum_{i=1}^{n} p_i \left(a x_i + b \right) \\
&= a \sum_{i=1}^{n} x_i p_i + b \underbrace{\sum_{i=1}^{n} p_i}_{=1} \\
&= aE[X] + b.
\end{aligned}
$$

Now, consider another r.v. defined by $Z = X - E[X]$ (which assumes values 1 with probability $p_\uparrow = \frac{1}{3}$ and -0.5 with $p_\downarrow = \frac{2}{3}$ in the above example), namely the deviation of X from its mean. The expected value of Z is

$$
E[Z] = E[X - E[X]] = E[X] - E[X] = 0 \qquad \forall \, X \, .
$$

Also, consider the r.v. $Z^2 = (X - E[X])^2$ (which assumes values 1 with probability $p_\uparrow = \frac{1}{3}$ and 0.25 $p_\downarrow = \frac{2}{3}$ in the above example), namely the squared deviation of X from its mean. The expected value of Z^2 is

$$
\begin{aligned}
E[Z^2] &= E[(X - E[X])^2] \\
&= p_\uparrow \left(x_\uparrow - E[X] \right)^2 + p_\downarrow \left(x_\downarrow - E[X] \right)^2 \\
&= \frac{1}{3} 1 + 0.25 \frac{2}{3} = \frac{1}{2} \neq 0 \qquad\qquad\qquad (2.5)
\end{aligned}
$$

and it represents the variance of X. We can now introduce the definition of variance of a random variable.

Definition 86 (Variance) *Let X be a random variable. The variance of X is defined as $Var[X] = E[(X - E[X])^2]$, namely the expected value of the squared deviation of X from its mean.*

Theorem 87 *Let X be a random variable. Then*

$$
Var[X] = \sigma_X^2 = E[X^2] - E[X]^2.
$$

Proof.

$$
\begin{aligned}
\sigma_X^2 &= E[(X - E[X])^2] = E[X^2 - 2XE[X] + E[X]^2] \\
&= E[X^2] - 2E[X]E[X] + E\left[E[X]^2\right] \\
&= E[X^2] - 2E[X]^2 + E[X]^2 \\
&= E[X^2] - E[X]^2 \, .
\end{aligned}
$$

∎

73

Example 88 *Let us continue from the example described at the beginning of this section:*

$$X = \begin{cases} 2 & \text{with} \quad p_\uparrow = \frac{1}{3} \\ 0.5 & \text{with} \quad p_\downarrow = \frac{2}{3} \end{cases}$$

Then $E[X] = 1$ *and*

$$\begin{aligned} E[X^2] &= p_\uparrow X(\omega_\uparrow)^2 + p_\downarrow X(\omega_\downarrow)^2 = \\ &= \frac{1}{3}4 + \frac{2}{3}\frac{1}{4} = \frac{3}{2}. \end{aligned}$$

Using Theorem 87 we have that $\sigma_X^2 = \frac{3}{2} - 1 = \frac{1}{2}$ *as in (2.5).*

Remark 89 (Variance of an affine transformation of a r.v.) *Give a random variable* X, *if we consider an affine transformation* $Z = aX + b$ *with* $a, b \in \mathbb{R}$, *it is straightforward to show that variance is not a linear operator:*

$$\begin{aligned} Var[Z] &= Var[aX + b] \\ &= E[(aX + b - aE[X] - b)^2] \\ &= E[a^2 (X - E[X])^2] \\ &= a^2 Var[X]. \end{aligned}$$

Definition 90 (Volatility) *Let* X *be a random variable. The standard deviation of* X *(called volatility in finance world) is defined as* $\sigma_X = \sqrt{\sigma_X^2}$.

Definition 91 (Expected value of a function of a r.v.) *Let* X *be a discrete random variable. We have*

$$E[g(X)] = \sum_{i=1}^{n} p_i g(x_i).$$

2.3 Probability distributions

As discussed in the previous section, to investigate a generic random variable X it is necessary to know not only the values assumed by X, but also the probabilities that these values will be realized. To specify such probabilities we introduce the concept of probability distribution of a random variable. A way to specify a random variable is by means of its probability density function (pdf). As shown in Example 93, in the case of a discrete r.v. the pdf is a function that provides the probabilities that a discrete random variable assumes specific values (see Fig. 2.3, left side).

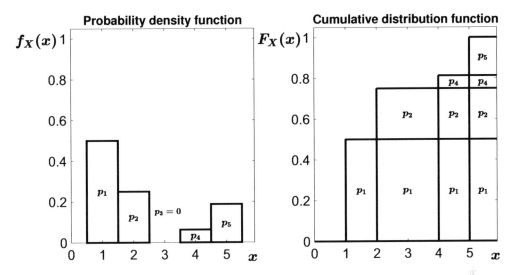

Figure 2.3: Pdf and cdf of the r.v. described in Example 93

Another equivalent way to describe a r.v. is by means of its cumulative distribution function (cdf) which is defined as follows.

Definition 92 (Cumulative distribution function) *Let X be a random variable. We define the cumulative distribution function of X as*

$$F_X(x) = P(X \leq x) \qquad where \quad x \in \mathbb{R}.$$

Example 93 *Let the outcomes of a r.v. X be $\{2, 4, 1, 5\}$ with probabilities $\{\frac{1}{4}, \frac{1}{16}, \frac{1}{2}, \frac{3}{16}\}$, as exhibited in Fig. 2.3 (left side). Then, using Definition 92 the cumulative distribution function $F_X(x)$ can be plotted as in Fig. 2.3 (right side).*

Remark 94 *From Example 93 and Fig. 2.3 we can intuitively claim that*

- $\lim\limits_{x \to -\infty} F_X(x) = 0$;

- $\lim\limits_{x \to +\infty} F_X(x) = 1$;

- $F_X(x)$ *can not be a continuous function.*

In the following exercise, we propose how to construct in practice the cdf of a discrete random variable.

2.4 Continuous random variables

In this section, we see how to characterize a continuous r.v. that is defined on a space of events with an infinite number of elements, as shown in the following example.

Example 96 (Elementary event: daily cash) *Assume that money is infinitely divisible. The elementary event* ω, *that represents the daily income of a shop, is any non-negative real number, namely* $\omega \in \Omega = [0, +\infty) = \mathbb{R}_+^0$. *Let* X *be the daily gain, where* $X : \Omega \to \mathbb{R}$. *Furthermore, suppose that fixed costs are* 10€, *then* $X(\omega) \in [-10, +\infty)$.

Remark 97 *In the case of a random variable* X *defined on a space of events* $\Omega = \{\omega_1, \omega_2, \cdots, \omega_N\}$ *with a finite number of elements, we have that* $P(\omega = \omega_k) = P(X(\omega) = X(\omega_k) = x_k) = p_k \geq 0$, *namely the probability that a discrete r.v. assumes a specific value could be different from zero. A different situation occurs in the case of a r.v. that can assume an infinite number of realizations. For instance, suppose that* $\Omega = [a, b] \subset \mathbb{R}$. *By definition we have* $P(\omega \in \Omega) = 1$. *Now, consider the probability of occurrence of an elementary event* $\tilde{\omega}$, $P(\omega = \tilde{\omega}) = P(X(\omega) = X(\tilde{\omega}) = \tilde{x})$. *Since the space of events* $\Omega = [a, b]$, *if* $P(X(\omega) = \tilde{x}) \neq 0$, *then* $P(\omega \in \Omega)$ *would be* $+\infty$, *contradicting the property that* $P(\Omega) = 1$ *as described in Definition 65. Thus, the probability that a continuous r.v.* X *assumes a specific value* $\tilde{x}$ *is zero, namely* $P(X(\omega) = \tilde{x}) = 0$. *In other words, we can say that the probability that* X *assumes a single value within an infinite set, is infinitesimally small, which statistically is equivalent to zero. Indeed, a coherent description for a continuous r.v. can be done by assigning a positive probability only for events that include infinitely many outcomes, such as follows*

$$P(\omega \in [\omega_1, \omega_2]) \geq 0 \qquad \forall \, \omega_1, \omega_2 \in \Omega \, .$$

	discrete r.v.	continuous r.v.
	$p_i \geq 0 \quad \forall i$	$f_X(x) \geq 0 \quad \forall\, x \in \mathbb{R}$
	$P(X \in (a, b]) = \sum_{a < x_i \leq b} p(x_i)$	$P(X \in (a, b]) = \int_a^b f_X(x)dx$
	$\sum_i p_i = 1$	$\int_{-\infty}^{\infty} f_X(x)dx = 1$
$E[X]$	$\sum_i x_i\, p_i$	$\int_{-\infty}^{+\infty} x\, f_X(x)dx$
$Var[X]$	$\sum_i (x_i - E[X])^2\, p_i$	$\int_{-\infty}^{+\infty} (x - E[X])^2\, f_X(x)dx$
$E[g(X)]$	$\sum_i g(x_i)\, p_i$	$\int_{-\infty}^{+\infty} g(x)\, f_X(x)dx$

Table 2.2: Comparison between continuous and discrete random variables

A random variable X is said to be *absolutely* continuous if there exists a function $f_X(x)$ such that

$$P(X \in [a, b]) = \int_a^b f_X(x)dx\,, \qquad (2.6)$$

where $f_X(x)$ is called the probability density function of X.
In the case of a discrete r.v. we have

$$P(X \in (a, b]) = \sum_{a < x_i \leq b} P(X = x_i)\,.$$

With some abuse of the notation, we can say that $f_X(x)$ has the same role of the $p_i = P(X = x_i)$ for the discrete r.v. as follows

$$\sum_i \underbrace{()}_{g(x_i)} p_i \rightarrow \int_{-\infty}^{+\infty} \underbrace{()}_{g(x)} f_X(x)dx\,.$$

In Table 2.2 we summarize the main similarities and differences between probabilities distributions of continuous and discrete random variables, and we report the mathematical formulation of the expected value, variance and expectation of a generic function of discrete and continuous random variables.

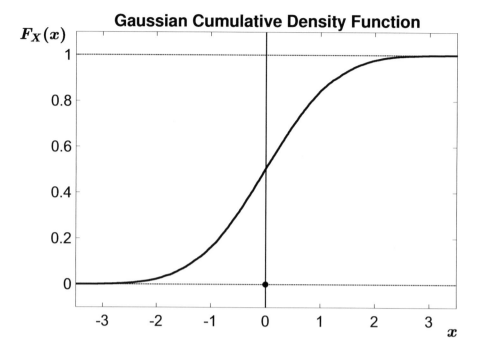

Figure 2.4: Cdf of a Standard Normal r.v.

Note that, using Expression (2.6) it is straightforward to verify what we claim in Remark 97 about continuous random variables

$$P(X = x) = \lim_{h \to 0} P(x \le X \le x + h)$$

$$= \lim_{h \to 0} \int_x^{x+h} f_X(u)du = 0 \,.$$

Definition 98 (CDF of a continuous r.v.) *Let X be a continuous random variable. We define its cumulative distribution function as*

$$F_X(x) = P(X \le x) = P(X \in (-\infty, x]) = \int_{-\infty}^x f_X(t)dt \,. \qquad (2.7)$$

An example of cdf of a continuous r.v. is reported in Fig. 2.4.

Continuing from Remark 94 and using Definition 98, it is interesting to observe that

- $\lim_{x \to -\infty} F_X(x) = 0$.

  Indeed, $\lim_{x \to -\infty} F_X(x) = \lim_{x \to -\infty} \int_x^x f_X(t)dt = 0$.

- $\lim\limits_{x \to +\infty} F_X(x) = 1$.

  Indeed, $\lim\limits_{x \to +\infty} F_X(x) = P(X \in (-\infty, +\infty)) = \int_{-\infty}^{+\infty} f_X(t)dt = 1$.

- $F_X(x)$ is a non-decreasing function of x.

  Claiming that $F_X(x)$ is a non-decreasing function of x means that if $x_1 \le x_2$, then $F_X(x_1) \le F_X(x_2)$. Now, from Definition 98 we have $F_X(x_2) = \int_{-\infty}^{x_2} f_X(t)dt = \int_{-\infty}^{x_1} f_X(t)dt + \int_{x_1}^{x_2} f_X(t)dt$. Since $\int_{x_1}^{x_2} f_X(t)dt = P(X \in (x_1, x_2)) \ge 0$, $F_X(x_2) - F_X(x_1) \ge 0$.

- $P(X \in [a,b]) = \int_a^b f_X(x)dx = F_X(b) - F_X(a)$.

  Indeed, $\int_{-\infty}^{b} f_X(x)dx = \int_{-\infty}^{a} f_X(x)dx + \int_a^b f_X(x)dx$. Thus, $\int_a^b f_X(x)dx = \int_{-\infty}^{b} f_X(x)dx - \int_{-\infty}^{a} f_X(x)dx = F_X(b) - F_X(a)$. Geometrically, the probability that the r.v. X can assume values in the interval $[a,b]$ represents the value of the gray area in Fig. 2.5.

- If X is an absolutely continuous r.v., then $f_X(x) = \dfrac{dF_X(x)}{dx}$.

  This results is similar to the fundamental theorem of calculus (also called Torricelli-Barrow theorem).

  Let us consider the probability that the r.v. X can assume values in the interval $[x, x + \Delta x]$ (where the infinitesimal interval $\Delta x \to 0^+$), namely

  $$P(X \in [x, x + \Delta x]) = \int_x^{x+\Delta x} f_X(x)dx \,.$$

  With some abuse of the notation, we can say that $\int_x^{x+\Delta x} f_X(x)dx \approx f_X(x)\Delta x$, thus $P(X \in [x, x + \Delta x]) = F_X(x + \Delta x) - F_X(x) = f_X(x)\Delta x$. This implies that $f_X(x) = \lim\limits_{\Delta x \to 0^+} \dfrac{F_X(x + \Delta x) - F_X(x)}{\Delta x} = \dfrac{dF_X(x)}{dx}$.

In the following exercise, we propose how to compute and draw the graph of an empirical probability density function of a random variable.

Exercise 99 *Build a function* F_Empirical_pdf *that plots the empirical probability density function of a time series* $y = [y_1, \ldots, y_n]$.
Test the Function F_Empirical_pdf *using as input the vector of random numbers* Y=randn(1000,1).

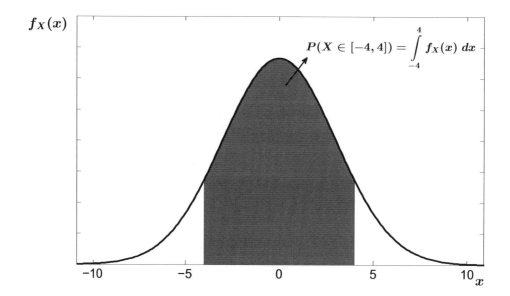

Figure 2.5: Probability that a r.v. can assume values into a generic interval

Hint: see the built-in functions `hist` *and* `bar` *to draw the graph of a pdf.*

Sol.: See Function `F_Empirical_pdf`.

2.5 Higher-order moments and synthetic indices of a distribution

In Sections 2.2 and 2.4, we defined and computed the expected value and the variance of a random variable X, namely $E[X]$ and $Var[X]$, respectively. Thus, extending the concept to any power k of a r.v. X we can define the raw moment of order k of X as follows:

$$\mu^{(k)} = E[X^k]$$

where $E[X^k] = \sum_{i=1}^{n} p_i X(\omega_i)^k = \sum_{i=1}^{n} p_i x_i^k$ for a discrete random variable, and $E[X^k] = \int_{-\infty}^{+\infty} x^k f_X(x)dx$ for a continuous random variable. The central moment of order k is defined by using the r.v. $X - E[X]$, namely

$$\bar{\mu}^{(k)} = E[(X - E[X])^k]$$

The first two synthetic indices used to investigate the shape of a distribution are the mean, i.e., $\mu^{(1)} = E[X^1]$ that represents the *location* of a distribution, and the variance, i.e., $\bar{\mu}^{(2)} = E[(X - E[X])^2] = \sigma_X^2$ that measures the *dispersion* of a distribution.

80

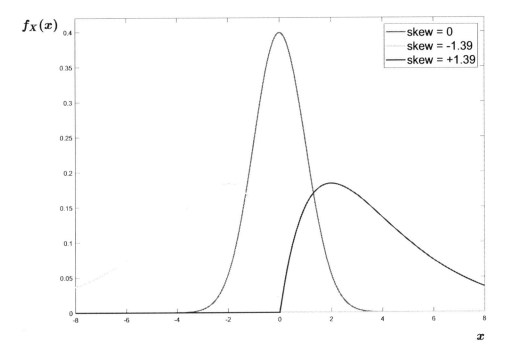

Figure 2.6: Example of pdfs with different values of skewness.

The central moment of order $k = 3$ describes the property of *skewness* (asymmetry) and is defined as

$$Sk[X] = \frac{\bar{\mu}^{(3)}}{\sigma_X^3} = E\left[\left(\frac{X - E[X]}{\sigma_X}\right)^3\right]$$

If the *skewness* of X approaches 0, the distribution tends to be symmetric with respect to the mean μ; if it is positive (negative), the tail is most marked on the right (left) side (see Fig. 2.6).
The *kurtosis* is defined by the central moment of order $k = 4$ as follows:

$$Ku[X] = \frac{\bar{\mu}^{(4)}}{\sigma_X^4} = E\left[\left(\frac{X - E[X]}{\sigma_X}\right)^4\right]$$

If the *kurtosis* approaches to 3, the shape of the distribution is similar to the normal distribution; if it is greater than 3, then the peak is tighter than that of the normal, and extreme events are more likely. On the other hand, if the *kurtosis* is less than 3, then the peak is wider than that of the normal and extreme events are less likely (see Fig. 2.7).

81

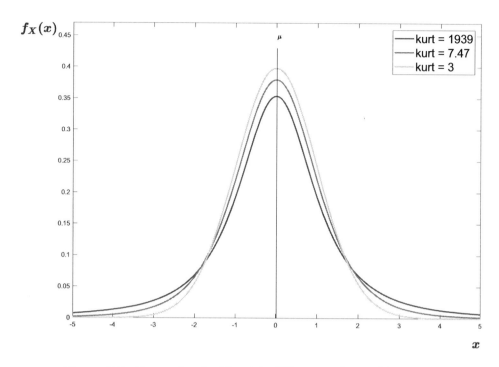

Figure 2.7: Example of pdfs with different values of kurtosis.

Exercise 100 (Synthetic indices of a distribution)
Write a Function (named F_SyntheticIndices) that computes the mean, the variance, the skewness and the kurtosis of a time series $y = [y_1, \ldots, y_n]$. In a Script, test the Function on a $n \times 1$ vector y of random numbers generated from a standard normal distribution, using the built-in function randn.

Sol.: See Function F_SyntheticIndices and Script S_SyntheticIndices.

2.6   Some probability distributions

In this section, we deal with some of the most frequently used probability distributions in finance: *Uniform, Normal, Log-normal, Chi-square* and *Student-t*. We characterize each random variable specifying its pdf, cdf and summary statistics. In addition, for each distribution we propose an exercise solved by using MATLAB.

2.6.1 Uniform distribution

Consider a random variable X uniformly distributed on a range of real numbers $[a, b]$, namely $X \sim U([a, b])$. The probability density function of X is

$$f_X(x) = \begin{cases} 0 & \text{for} \quad x \leq a \\ \dfrac{1}{b-a} & \text{for} \quad a < x \leq b \\ 0 & \text{for} \quad x > b \end{cases} \qquad (2.8)$$

Furthermore, the uniform random variable X has the following cumulative distribution function:

$$F_X(x) = \begin{cases} 0 & \text{for} \quad x \leq a \\ \dfrac{x-a}{b-a} & \text{for} \quad a < x \leq b \\ 1 & \text{for} \quad x > b \end{cases} \qquad (2.9)$$

Fig. 2.8 shows theoretical and empirical pdfs (at the top), and theoretical and empirical cdfs (at the bottom) of a standard uniform random variable.

Example 101 *Let X be an r.v. with $f_X(x) = K$ (constant) for $x \in (a, b]$, and $f_X(x) = 0$ otherwise. Show that X has a pdf and a cdf as in Expressions (2.8) and (2.9), respectively.*
Applying the property that $P(X \in (-\infty, +\infty)) = 1$, we have

$$\begin{aligned} P(X \in (-\infty, +\infty)) &= 1 \\ \Rightarrow F_X(+\infty) &= \int_{-\infty}^{+\infty} f_X(x)dx = 1 \\ \Rightarrow \int_a^b K\,dx &= K(b-a) = 1 \\ \Rightarrow K &= \frac{1}{b-a}. \end{aligned}$$

Furthermore, considering Definition 2.7 and that $f_X(x) = 0$ for $x \leq a$, we can write

$$F_X(x) = \int_{-\infty}^x f_X(t)dt = \frac{1}{b-a}\int_a^x dt = \frac{x-a}{b-a}.$$

The synthetic indices of the uniform distribution are:

$$E[X] = a + \frac{b-a}{2} \qquad (2.10)$$

Standard Uniform Distribution

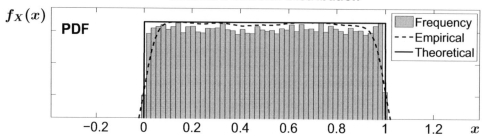

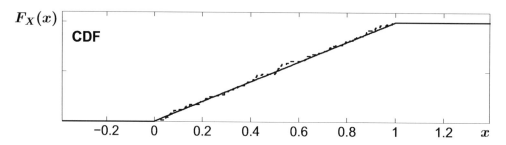

Figure 2.8: Pdf and cdf of a standard uniform r.v.

$$Var[X] = \frac{(b-a)^2}{12} \tag{2.11}$$

$$Sk[X] = 0 \tag{2.12}$$

$$Ku[X] = \frac{9}{5}. \tag{2.13}$$

Example 102 *Using the definitions of the expected value and of the variance for a continuous r.v. in Table 2.2, the following steps show how to obtain Expressions (2.10) and (2.11), respectively. In the case of a uniform r.v. its pdf is given by (2.8), then we have*

$$E[X] = \int_{-\infty}^{+\infty} x f_X(x) dx =$$

$$= \frac{1}{b-a} \int_a^b x dx$$

$$= \frac{1}{b-a} \left[\frac{x^2}{2} \right]_a^b$$

$$= \frac{b+a}{2} = a + \frac{b-a}{2} .$$

Furthermore, we can write that

$$E\left[X^2\right] = \int_{-\infty}^{+\infty} x^2 f_X(x) dx = \frac{1}{b-a} \int_a^b x^2 dx$$

$$= \frac{1}{b-a} \left[\frac{x^3}{3} \right]_a^b$$

$$= \frac{1}{3} \frac{1}{b-a} \left(b^3 - a^3 \right)$$

$$= \frac{1}{3} \frac{1}{b-a} (b-a) \left(b^2 + ab + a^2 \right) .$$

Thus using Theorem 87 we can easily obtain the variance of $X \sim U([a,b])$

$$\sigma_X^2 = E\left[X^2\right] - (E[X])^2$$

$$= \frac{1}{3} \left(b^2 + ab + a^2 \right) - \frac{b^2 + 2ab + a^2}{4}$$

$$= \frac{b^2 + a^2 - 2ab}{12}$$

$$= \frac{(b-a)^2}{12} .$$

Exercise 103 (Uniform distribution)
Generate an $n \times 1$ vector x of random numbers generated from a uniform distribution on the interval $[a,b]$, where $a = 2$ and $b = 7$. Using the functions F_Empirical_pdf and F_Empirical_cdf, plot the empirical pdf and cdf of the outcomes x. Note that the built-in function rand *generates uniformly distributed pseudorandom numbers on the interval $[0,1]$. Therefore, values from the uniform distribution on the interval $[a,b]$ can be obtained*

considering the following affine transformation x = a+(b-a).*rand(n,1).
Furthermore, add the analytical pdf and cdf to the previous plots using
the built-in functions pdf *and* cdf. *In addition, compute the mean, the*
variance, the skewness and the kurtosis numerically, using the Function
F_SyntheticIndices *of Ex. 100, and compare them with the analytical*
expressions (2.10), (2.11), (2.12) and (2.13).

Sol.: See Script S_UnEmpPdfCdf.

2.6.2 Normal distribution

Let us denote a normal random variable by $X \sim N(\mu, \sigma^2)$ which has expected
value μ and variance σ^2. When $\mu = 0$ and $\sigma = 1$, we obtain a standard normal
r.v. $Z \sim N(0, 1)$. The pdf of X is as follows

$$f_X(x) = \frac{1}{\sqrt{2\pi\sigma^2}} e^{-\frac{(x-\mu)^2}{2\sigma^2}}. \tag{2.14}$$

Equivalently, we can characterize the normal random variable X by means of
its cumulative distribution function

$$F_X(x) = \frac{1}{\sqrt{2\pi\sigma^2}} \int_{-\infty}^{x} e^{-\frac{(t-\mu)^2}{2\sigma^2}} dt = \tag{2.15}$$

$$= \frac{1}{2}\left[1 + \mathrm{erf}\left(\frac{x-\mu}{\sqrt{2\sigma^2}}\right)\right], \tag{2.16}$$

where

$$\mathrm{erf}(z) = \frac{2}{\sqrt{\pi}} \int_0^z e^{-t^2} dt \tag{2.17}$$

is the Gauss error function. Fig. 2.9 shows theoretical and empirical pdfs (at the
top), and theoretical and empirical cdfs (at the bottom) of a standard normal
random variable.

Remark 104 (Error Function) *The error function (2.17) is a special func-*
tion largely used in Probability and Statistics. As shown in Fig. 2.10, the error
function has two horizontal asymptotes in 1 and -1. More specifically,

$$\mathrm{erf}(+\infty) = \frac{2}{\sqrt{\pi}} \int_0^{+\infty} e^{-t^2} dt = +1 \tag{2.18}$$

$$\mathrm{erf}(-\infty) = \frac{2}{\sqrt{\pi}} \int_0^{-\infty} e^{-t^2} dt = -1. \tag{2.19}$$

Expressions (2.18) and (2.19) can be found by using the Gauss integral
$\int_{-\infty}^{+\infty} e^{-t^2} dt = \sqrt{\pi}$. *Indeed, since* $g(t) = e^{-t^2}$ *is an even function, namely*

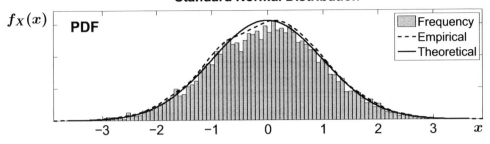

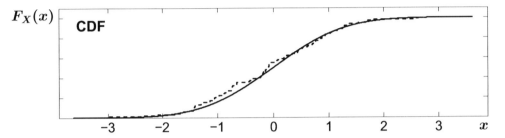

Figure 2.9: Pdf and cdf of a standard normal r.v.

$g(-t) = g(t)$, *we can write that*

$$\int_{-\infty}^{+\infty} e^{-t^2} dt = \int_{-\infty}^{0} g(t)dt + \int_{0}^{+\infty} g(t)dt$$

$$= -\int_{+\infty}^{0} g(-t)dt + \int_{0}^{+\infty} g(t)dt$$

$$= 2\int_{0}^{+\infty} g(t)dt = 2\int_{-\infty}^{0} g(t)dt$$

Thus, $\int_{0}^{+\infty} e^{-t^2} dt = \int_{-\infty}^{0} e^{-t^2} dt = \dfrac{\sqrt{\pi}}{2}.$

The values of the error function can be obtained by means of the Taylor expansion of the integrand of (2.17) and by integrating as follows:

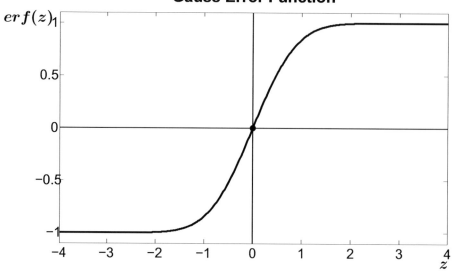

Figure 2.10: Graph of the error function (2.17)

$$
\begin{aligned}
\mathrm{erf}(z) &= \frac{2}{\sqrt{\pi}} \int_0^z e^{-t^2} dt \\
&= \frac{2}{\sqrt{\pi}} \int_0^z \left(\sum_{k=0}^{+\infty} (-1)^k \frac{t^{2k}}{k!} \right) dt \\
&= \frac{2}{\sqrt{\pi}} \sum_{k=0}^{+\infty} \frac{(-1)^k}{k!} \int_0^z t^{2k} dt \\
&= \frac{2}{\sqrt{\pi}} \sum_{k=0}^{+\infty} \frac{(-1)^k}{k!} \frac{z^{2k+1}}{2k+1}.
\end{aligned}
$$

Clearly, the greater the number of terms considered in the summation, the better the precision of the value of the error function obtained.

Example 105 (Cdf of a standard normal and the error function)
In the case of a standard normal random variable $Z \sim N(0,1)$, verify that the cdf can be expressed in terms of the error function, as in (2.16). Note that usually, when $\mu = 0$ and $\sigma = 1$, the cumulative density function (2.15)

is denoted by $\Phi(z)$.

$$\begin{aligned}
\Phi(z) &= \frac{1}{\sqrt{2\pi}} \int_{-\infty}^{z} e^{-\frac{t^2}{2}} dt \\
&= \frac{1}{\sqrt{2\pi}} \left[\int_{-\infty}^{0} e^{-\frac{t^2}{2}} dt + \int_{0}^{z} e^{-\frac{t^2}{2}} dt \right] \\
&= \frac{1}{\sqrt{2\pi}} \left[-\int_{0}^{-\infty} e^{-\frac{t^2}{2}} dt + \int_{0}^{z} e^{-\frac{t^2}{2}} dt \right].
\end{aligned}$$

Now, let us consider $u = \dfrac{t}{\sqrt{2}}$, therefore $du = \dfrac{1}{\sqrt{2}} dt \Rightarrow dt = \sqrt{2} du$. Thus,

$$\begin{aligned}
\Phi(z) &= \frac{1}{\sqrt{2\pi}} \left[-\sqrt{2} \int_{0}^{-\infty} e^{-u^2} du + \sqrt{2} \int_{0}^{\frac{z}{\sqrt{2}}} e^{-u^2} du \right] \\
&= \frac{1}{2} \left[-\frac{2}{\sqrt{\pi}} \int_{0}^{-\infty} e^{-u^2} du + \frac{2}{\sqrt{\pi}} \int_{0}^{\frac{z}{\sqrt{2}}} e^{-u^2} du \right] \\
&= \frac{1}{2} \left[-\operatorname{erf}(-\infty) + \operatorname{erf}\left(\frac{z}{\sqrt{2}}\right) \right] = \frac{1}{2} \left[1 + \operatorname{erf}\left(\frac{z}{\sqrt{2}}\right) \right].
\end{aligned}$$

The synthetic indices used to characterize the properties of the normal distribution are expected value, variance, skewness and kurtosis:

$$E[X] = \mu \tag{2.20}$$

$$Var[X] = \sigma^2 \tag{2.21}$$

$$Sk[X] = 0 \tag{2.22}$$

$$Ku[X] = 3. \tag{2.23}$$

Example 106 *Using the definitions of the expected value and of the variance for a continuous r.v. in Table 2.2, the following steps show how to obtain Expressions (2.20) and (2.21), respectively.*

$$E(x) = \frac{1}{\sqrt{2\pi\sigma^2}} \int_{-\infty}^{+\infty} x e^{-\frac{(x-\mu)^2}{2\sigma^2}} dx.$$

Now, let us consider $t = \dfrac{x - \mu}{\sigma}$, *therefore* $x = \sigma t + \mu \Rightarrow dx = \sigma dt$. *Thus,*

$$
\begin{aligned}
E(x) &= \frac{\sigma}{\sqrt{2\pi\sigma^2}} \int_{-\infty}^{+\infty} (\sigma t + \mu) e^{-\frac{t^2}{2}} dt \\
&= \frac{1}{\sqrt{2\pi}} \left[\sigma \int_{-\infty}^{+\infty} te^{-\frac{t^2}{2}} dt + \mu \int_{-\infty}^{+\infty} e^{-\frac{t^2}{2}} dt \right].
\end{aligned}
$$

Note that $\int_{-\infty}^{+\infty} te^{-\frac{t^2}{2}} dt = 0$. *Indeed, since* $h(t) = te^{-\frac{t^2}{2}}$ *is an odd function, namely* $h(-t) = -h(t)$, *we can write that*

$$
\begin{aligned}
\int_{-\infty}^{+\infty} te^{-\frac{t^2}{2}} dt &= \int_{-\infty}^{0} h(t)dt + \int_{0}^{+\infty} h(t)dt \\
&= -\int_{+\infty}^{0} h(-u)du + \int_{0}^{+\infty} h(t)dt \\
&= \int_{0}^{+\infty} h(-u)du + \int_{0}^{+\infty} h(t)dt \\
&= -\int_{0}^{+\infty} h(u)du + \int_{0}^{+\infty} h(t)dt = 0 \,,
\end{aligned}
$$

where we set $u = -t \Rightarrow du = -dt$. *Then,*

$$
E[X] = \mu \frac{1}{\sqrt{2\pi}} \int_{-\infty}^{+\infty} e^{-\frac{t^2}{2}} dt \,.
$$

Considering $y = \dfrac{t}{\sqrt{2}}$, *therefore* $dy = \dfrac{1}{\sqrt{2}} dt \Rightarrow dt = \sqrt{2} dy$, *we obtain*

$$
E[X] = \mu \frac{1}{\sqrt{\pi}} \overbrace{\int_{-\infty}^{+\infty} e^{-y^2} dy}^{=\sqrt{\pi}} = \mu \,.
$$

Starting from Definition 86, we have

$$
Var[X] = E[(X - E[X])^2] = \frac{1}{\sqrt{2\pi\sigma^2}} \int_{-\infty}^{+\infty} (x - \mu)^2 e^{-\frac{(x-\mu)^2}{2\sigma^2}} dx \,.
$$

Consider $t = \frac{x-\mu}{\sigma}$, *therefore* $x = \sigma t + \mu \Rightarrow dx = \sigma dt$. *Thus*

$$\begin{aligned} Var[X] &= \frac{1}{\sqrt{2\pi\sigma^2}} \int_{-\infty}^{+\infty} \sigma^2 t^2 e^{-\frac{t^2}{2}} \sigma dt \\ &= \frac{\sigma^2}{\sqrt{2\pi}} \int_{-\infty}^{+\infty} t^2 e^{-\frac{t^2}{2}} dt. \end{aligned}$$

Now, let us integrate by part, namely $\int f'(t)g(t)dt = f(t)g(t) - \int f(t)g'(t)dt$, *where* $g(t) = t$ *and* $f'(t) = te^{-\frac{t^2}{2}}$, *and, therefore,* $g'(t) = 1$ *and* $f(t) = -e^{-\frac{t^2}{2}}$. *Then, we have*

$$\begin{aligned} Var[X] &= \frac{\sigma^2}{\sqrt{2\pi}} \left(-te^{-\frac{t^2}{2}} \Big|_{-\infty}^{+\infty} - \int_{-\infty}^{+\infty} -e^{-\frac{t^2}{2}} dt \right) \\ &= \frac{\sigma^2}{\sqrt{2\pi}} \int_{-\infty}^{+\infty} e^{-\frac{t^2}{2}} dt = \frac{\sigma^2}{\sqrt{2\pi}} \sqrt{2\pi} = \sigma^2. \end{aligned}$$

Note that since $\lim_{t \to \infty} \frac{-t}{e^{t^2/2}} = 0$, $-te^{-\frac{t^2}{2}} \Big|_{-\infty}^{+\infty} = 0$, *and that* $\int_{-\infty}^{+\infty} e^{-\frac{t^2}{2}} dt = \sqrt{2\pi}$ *which follows from the Gauss integral.*

Exercise 107 (Normal distribution) *Create an* $n \times 1$ *vector* x *of random numbers generated from a Normal distribution with mean* $\mu = 1$ *and standard deviation* $\sigma = 2$. *Using the Functions* F_Empirical_pdf *and* F_Empirical_cdf, *plot the empirical pdf and cdf of the outcomes* x. *Note that the built-in function* randn *generates normally distributed pseudorandom numbers with* $\mu = 0$ *and* $\sigma = 1$ *(standard normal). However, values from the normal distribution for any* μ *and* σ *can be obtained considering the following affine transformation* x = mu + sigma * randn(n,1), *where* n *is the number of simulations. Furthermore, add the analytical pdf and cdf to the previous plots using the built-in functions* pdf *and* cdf. *In addition, compute the mean, the variance, the skewness and the kurtosis numerically, using the Function* F_SyntheticIndices *of Ex. 100, and compare them with the analytical expressions (2.20), (2.21), (2.22) and (2.23).*

Sol.: See Script S_NormEmpPdfCdf.

2.6.3   Log-normal distribution

A possible way to represent a random variable that can assume only positive values is to use a Log-Normal random variable. An r.v. X has a log-normal distribution if $\ln(X) \sim N(\mu, \sigma^2)$. It is possible to show that the pdf of X is as follows

$$f_X(x) = \begin{cases} 0 & \text{for} \quad x \leq 0 \\ \dfrac{1}{x\sqrt{2\pi\sigma^2}} e^{-\frac{(\ln x - \mu)^2}{2\sigma^2}} & \text{for} \quad x > 0 \end{cases} \tag{2.24}$$

Proof. Let us start from the cdf of a normal r.v. $Y = \ln(X) \sim N(\mu, \sigma^2)$.

$$P(Y \leq y) = P(\ln(X) \leq y) = \int_{-\infty}^{y} \frac{1}{\sqrt{2\pi\sigma^2}} e^{-\frac{(u-\mu)^2}{2\sigma^2}} \, du \, .$$

Let $u = \ln t \Rightarrow du = \frac{dt}{t}$. Furthermore, since $t = e^u$, the new interval of integration becomes $(0, e^y]$. Therefore, we have

$$P(\ln(X) \leq y) = \int_{0}^{e^y} \frac{1}{t\sqrt{2\pi\sigma^2}} e^{-\frac{(\ln t - \mu)^2}{2\sigma^2}} \, dt \, .$$

Let $y = \ln x$, or, equivalently, $x = e^y$. Thus

$$P(\ln(X) \leq \ln(x)) = \int_{0}^{x} \frac{1}{t\sqrt{2\pi\sigma^2}} e^{-\frac{(\ln t - \mu)^2}{2\sigma^2}} \, dt \, .$$

Now, since the logarithmic function is a strictly increasing function, $P(\ln(X) \leq \ln(x)) = P(X \leq x)$, namely

$$P(X \leq x) = \int_{0}^{x} \frac{1}{t\sqrt{2\pi\sigma^2}} e^{-\frac{(\ln t - \mu)^2}{2\sigma^2}} \, dt \tag{2.25}$$

$$= \int_{-\infty}^{x} f_X(t) \, dt \, , \tag{2.26}$$

where $f_X(t)$ is the pdf described in (2.24). ∎

Furthermore, the log-normal random variable X has the following cumulative distribution function:

$$F_X(x) = \int_{-\infty}^{x} f_X(t) \, dt =$$

$$= \frac{1}{2}\left[1 + \text{erf}\left(\frac{\ln x - \mu}{\sqrt{2\sigma^2}} \right) \right] \, .$$

Fig. 2.11 shows theoretical and empirical pdfs (at the top), and theoretical and empirical cdfs (at the bottom) of a log-normal random variable.

The synthetic indices used to characterize the properties of the log-normal distribution are:

$$E[X] = e^{\mu + \frac{\sigma^2}{2}} \tag{2.27}$$

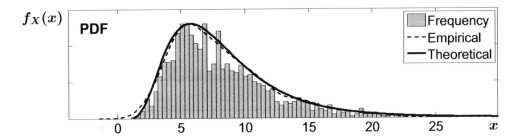

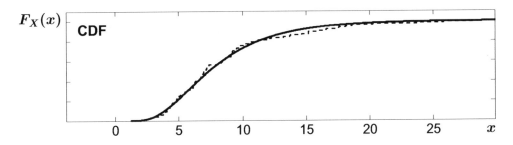

Figure 2.11: Pdf and cdf of a log-normal r.v.

$$Var[X] = \left(e^{\sigma^2} - 1\right)\left(e^{\mu + \frac{\sigma^2}{2}}\right)^2 \tag{2.28}$$

$$Sk[X] = \left(e^{\sigma^2} + 2\right)\left(e^{\sigma^2} - 1\right)^{\frac{1}{2}} \tag{2.29}$$

$$Ku[X] = e^{4\sigma^2} + 2e^{3\sigma^2} + 3e^{2\sigma^2} - 3. \tag{2.30}$$

Exercise 108 (Log-normal distribution) *Create an $n \times 1$ vector x of random numbers generated from a Log-normal distribution with parameters $\mu = 2$ and $\sigma = 0.5$. Using the Functions* F_Empirical_pdf *and* F_Empirical_cdf, *plot the empirical pdf and cdf of the outcomes x. Furthermore, add the analytical pdf and cdf to the previous plots using the built-in functions* pdf *and* cdf. *In addition, compute the mean, the variance, the skewness and the kurtosis numerically, using the Function* F_SyntheticIndices *of Ex. 100, and compare them with the analytical expressions (2.27), (2.28), (2.29) and (2.30).*

Sol.: See Script S_LogNormEmpPdfCdf.

Chi-Square Distribution

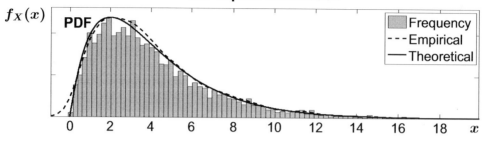

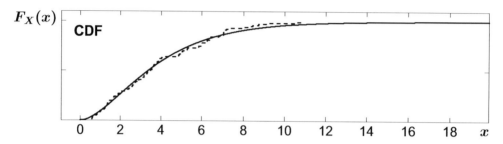

Figure 2.12: Pdf and cdf of a chi-square r.v.

2.6.4 Chi-square distribution

Let $Z_1, Z_2, \ldots, Z_\nu$ be ν independent and identically distributed (i.i.d.) random variables, where $Z_k \sim N(0,1)$ with $k = 1, 2, \ldots, \nu$. The random variable $X = \sum_{k=1}^{\nu} Z_k^2$ is characterized by a chi-square distribution with ν degrees of freedom. The pdf of a chi-square distributed r.v., namely more compactly $X \sim \chi_\nu^2$, is

$$f_X(x) = \begin{cases} 0 & \text{for} \quad x \le 0 \\ C_\nu e^{-\frac{x}{2}} x^{\frac{\nu}{2}-1} & \text{for} \quad x > 0 \end{cases} \tag{2.31}$$

where $C_\nu = \left(2^{\frac{\nu}{2}} \Gamma(\frac{\nu}{2})\right)^{-1}$ and $\Gamma(\frac{\nu}{2})$ is the Euler's Gamma function, which, for integer ν, has the following closed-form values

$$\Gamma\left(\frac{\nu}{2}\right) = \left(2^{\frac{\nu-1}{2}}\right)^{-1} (\nu - 2)(\nu - 4) \cdots \nu_0 \sqrt{\pi} \quad \text{with} \quad \nu_0 = \begin{cases} 1 & \text{if} \quad \nu \text{ is odd} \\ 2 & \text{if} \quad \nu \text{ is even} \end{cases}$$

Fig. 2.12 shows theoretical and empirical pdfs (at the top), and theoretical and empirical cdfs (at the bottom) of a chi-square random variable.

The synthetic indices used to characterize the properties of the Chi-square distribution are the following expected value, variance, skewness and kurtosis

$$E[X] = \nu \tag{2.32}$$

$$Var[X] = 2\nu \tag{2.33}$$

$$Sk[X] = \sqrt{\frac{8}{\nu}} \tag{2.34}$$

$$Ku[X] = \frac{12}{\nu} + 3. \tag{2.35}$$

Exercise 109 (Chi-square distribution) *Create an $n\times1$ vector x of random numbers generated from a Chi-square distribution with the degree of freedom $\nu = 6$. Using the Functions* F_Empirical_pdf *and* F_Empirical_cdf, *plot the empirical pdf and cdf of the outcomes x. Furthermore, add the analytical pdf and cdf to the previous plots using the built-in functions* pdf *and* cdf. *In addition, compute the mean, the variance, the skewness and the kurtosis numerically, using the Function* F_SyntheticIndices *of Ex. 100, and compare them with the analytical expressions (2.32), (2.33), (2.34) and (2.35).*

Sol.: See Script S_Chi2EmpPdfCdf.

2.6.5 Student-t distribution

Let $Z \sim N(0,1)$ and $Q \sim \chi_\nu^2$ a Chi-square random variable with ν degrees of freedom, then the standard Student-t is defined as $T = \dfrac{Z}{\sqrt{Q/\nu}}$. However, the standard Student-t distribution can be generalized using the following affine transformation $X = \mu + \sigma T \sim StudT(\mu, \sigma, \nu)$. The pdf of a generalized Student-t r.v. X is

$$f_X(x) = \frac{\Gamma(\frac{\nu+1}{2})}{\Gamma(\frac{\nu}{2})} \frac{1}{\sqrt{\nu\pi\sigma^2}} \left(1 + \frac{(x-\mu)^2}{\nu\sigma^2}\right)^{-\left(\frac{\nu+1}{2}\right)}, \tag{2.36}$$

where Γ is the Euler's Gamma function.

Fig. 2.13 shows theoretical and empirical pdfs (at the top), and theoretical and empirical cdfs (at the bottom) of a Student-t random variable.

The synthetic indices used to characterize the properties of the generalized

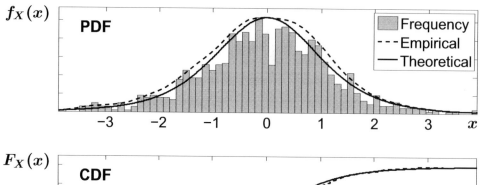

Figure 2.13: Pdf and cdf of a Student-t r.v.

Student-t distribution X are:

$$E[X] = \begin{cases} \text{undefined} & \text{for} \quad \nu = 1 \\ \mu & \text{for} \quad \nu > 1 \end{cases} \tag{2.37}$$

$$Var[X] = \begin{cases} \text{undefined} & \text{for} \quad \nu \leq 2 \\ \frac{\nu}{\nu-2}\sigma^2 & \text{for} \quad \nu > 2 \end{cases} \tag{2.38}$$

$$Sk[X] = \begin{cases} \text{undefined} & \text{for} \quad \nu \leq 3 \\ 0 & \text{for} \quad \nu > 3 \end{cases} \tag{2.39}$$

$$Ku[X] = \begin{cases} \text{undefined} & \text{for} \quad \nu \leq 4 \\ \frac{6}{\nu-4} + 3 & \text{for} \quad \nu > 4 \end{cases} \tag{2.40}$$

Exercise 110 (Student-t distribution) *Create an $(n \times 1)$ vector x of random numbers generated from a Student-t distribution with parameters $\mu = 0$, $\sigma = 1$, and with the degrees of freedom $\nu = 5$. Using the Functions* F_Empirical_pdf *and* F_Empirical_cdf, *plot the empirical pdf and cdf of the outcomes x. Furthermore, add the analytical pdf and cdf to the previous plots using the built-in functions* pdf *and* cdf. *In addition, compute*

96

the mean, the variance, the skewness and the kurtosis numerically, using the Function `F_SyntheticIndices` *of Ex. 100, and compare them with the analytical expressions (2.37), (2.38), (2.39) and (2.40).*

Sol.: See Script `S_StudTEmpPdfCdf`.

Chapter 3

Linear and Non-linear Programming

Several problems in Finance require the solution of an optimization problem. In other words, we are looking for the best value (a maximum or a minimum) of a function of one or more variables. We have a *constrained optimization* problem if such variables must belong to a given set, called the feasible region, and an *unconstrained optimization* problem otherwise. In Section 3.1 we present some basic concepts in Optimization. In Section 3.2.1 we discuss the most manageable mathematical optimization problem, where all functions that define the model are linear, and we introduce the specific tool of MATLAB to solve Linear Programming problems. Then, we describe more general problems, thus presenting the MATLAB optimization tool for Quadratic Programming (see Section 3.2.2) and for Nonlinear Programming (see Section 3.2.3). Finally, in Section 3.3 we deal with multi-objective optimization problems, describing how to reformulate them as single-objective optimization problems and constructing the efficient frontier.

3.1 General Framework

An optimization problem is a problem in which one wants to minimize or maximize a quantity represented by a real function, whose variables could be constrained to belong to a fixed set. Then, in general terms, given a non-empty set $C \subseteq \mathbb{R}^n$ and a function $f : C \to \mathbb{R}$, we could look for the solutions of the following optimization problems:

1. *Minimization*: $x^* \in C$ such that $f(x^*) \leq f(x) \quad \forall x \in C$;

2. *Maximization:* $\widehat{x} \in C$ such that $f(\widehat{x}) \geq f(x) \quad \forall x \in C$.

However, note that maximizing f on C corresponds to minimize $-f$ on the same set, more precisely

$$\max_{x \in C} f(x) = -\min_{x \in C} -f(x).$$

Thus, it is always possible to express an optimization problem in the following standard form

$$\begin{cases} \min_x \quad f(x) \\ \text{s.t.} \\ \qquad x \in C \end{cases} \tag{3.1}$$

where $x = (x_1, x_2, \ldots, x_n)$ represents the vector of the decisional variables, f is the objective function, and C is the feasible set, which is determined by the constraints of the problem. A point $x \in C$ is called a feasible solution.

An optimization problem is said to be *infeasible* if $C = \emptyset$, and it is said to be *unbounded* below (above) if $\forall m \, (M)$ there exists $x \in C$ such that $f(x) < m \, (f(x) > M)$. Furthermore, as mentioned earlier, Problem (3.1) admits optimal solutions, if there exist one or more points $x^* \in C$ such that $f(x^*) \leq f(x)$ for all $x \in C$.

In the case of a general nonlinear optimization problem, the feasible set C can be expressed by a system of equalities/inequalities that represents the constraints of the problem. More precisely, Problem (3.1) can be formulated as follows:

$$\begin{cases} \min_x \quad f(x_1, x_2, \ldots, x_n) \\ \text{s.t.} \\ \qquad g_1(x_1, x_2, \ldots, x_n) \leq b_1 \\ \qquad g_2(x_1, x_2, \ldots, x_n) \leq b_2 \\ \qquad \vdots \\ \qquad g_m(x_1, x_2, \ldots, x_n) \leq b_m \end{cases} \tag{3.2}$$

and, according to the type of the objective function and of the functions that represent the constraints, Problem (3.2) can be classified into a specific category of mathematical programming problems.

3.2 Optimization with MATLAB®

In this section we provide basic elements on constrained optimization, focusing on how to identify the main types of optimization problems covered in this book and how to solve them using MATLAB.

3.2.1 Linear Programming

At first, one can mention the class of Linear Programming (LP) problems, in which both the objective function and the constraints of Problem (3.2) are linear.

Remark 111 (Linear function) *A function $f : \mathbb{R}^n \to \mathbb{R}$ is linear if and only if the following properties are verified:*

1. *$\forall\, x', x'' \in \mathbb{R}^n \to f\,(x' + x'') = f\,(x') + f\,(x'')$ (additivity);*

2. *$\forall\, x' \in \mathbb{R}^n$ and $\lambda \in \mathbb{R} \to f\,(\lambda x') = \lambda f\,(x')$ (homogeneity of degree 1).*

In this case, $f(x) = f\,(x_1, x_2, \ldots, x_n)$ must be a polynomial function of x of degree one as follows:

$$f(x) = c_1 x_1 + c_2 x_2 + \ldots + c_n x_n = c^T x\,,$$

where $c \in \mathbb{R}^n$.

Thus, when the objective function and the constraints are linear functions, Problem (3.2) can be expressed as the following LP problem:

$$
\left\{
\begin{aligned}
&\min_{x} \quad c_1 x_1 + c_2 x_2 + \ldots + c_n x_n \\
&\text{s.t.} \\
&\qquad g_1(x) = a_{11} x_1 + a_{12} x_2 + \ldots + a_{1n} x_n \leq b_1 \\
&\qquad g_2(x) = a_{21} x_1 + a_{22} x_2 + \ldots + a_{2n} x_n \leq b_2 \\
&\qquad \vdots \\
&\qquad g_m(x) = a_{m1} x_1 + a_{m2} x_2 + \ldots + a_{mn} x_n \leq b_m
\end{aligned}
\right.
$$

More concisely, in matricial form we have

$$
\left\{
\begin{aligned}
&\min_{x} \quad c^T x \\
&\text{s.t.} \\
&\qquad Ax \leq b
\end{aligned}
\right.
\tag{3.3}
$$

where A is an $m \times n$ matrix and $b \in \mathbb{R}^m$. Linear Programming is one of the simplest types of problems with very efficient algorithms for its resolution. MATLAB provides some specific built-in functions for solving optimization problems, which are contained in the Optimization toolbox. Using the notations provided by the MATLAB Help, it is possible to express an LP problem in the following manner

$$
\left\{
\begin{aligned}
&\min \quad f^T x \\
&\text{s.t.} \\
&\qquad Ax \leq b \\
&\qquad A_{eq} x = b_{eq} \\
&\qquad l_b \leq x \leq u_b
\end{aligned}
\right.
\tag{3.4}
$$

where f, b, b_{eq}, x, l_b, and u_b are vectors, while A and A_{eq} are matrices. The first line of Problem (3.4) represents the objective function, while the second and the third ones describe the inequality and equality constraints, respectively. Furthermore, the fourth line indicates that the decision variables x must belong to the range specified by the lower (l_b) and upper (u_b) bounds. Often, these are called box constraints. To solve an LP problem, MATLAB Optimization toolbox provides the built-in function linprog, which has the following syntax:

```
[x,fval] = linprog(f,A,b,Aeq,beq,l_b,u_b) .
```

As one can note from the Help, l_b and u_b can be optional. Indeed, if these are not specified, the decision variables have not to satisfy box constraints. Furthermore, if no equality constraints exist, then one can impose Aeq=[ ] and beq=[ ], while if no inequality constraints exist, then one can fix A=[ ] and b=[ ]. Note also that the outputs of linprog, x and fval represent the optimal solution x^* and the optimal value $f^T x^*$ of Problem (3.4), respectively. The built-in function linprog includes four optimization algorithms (i.e., interior-point, dual-simplex, simplex, active-set) used to find x^*. For more details, see the MATLAB Help.

Exercise 112 *Solve the following LP problem:*

$$
\begin{cases}
\text{min} & -5x_1 - 4x_2 - 6x_3 \\
\text{s.t.} & \\
& x_1 - x_2 + x_3 \leq 20 \\
& 3x_1 + 2x_2 + 4x_3 \leq 42 \\
& -3x_1 - 2x_2 \geq -30 \\
& x_1, x_2, x_3 \geq 0
\end{cases}
$$

Sol.:$[x^* = (0, 15, 3); f(x^*) = -78]$. See Script S_Linprog.

3.2.2 Quadratic Programming

Problem (3.2) is said to be nonlinear if at least one of the functions f or g_i (with $i = 1, \ldots, m$) is not linear. A relevant special case is that where the objective function f is quadratic, i.e., it has the following form

$$
f(x) = x^T Q x + c^T x = \sum_{i=1}^{n} \sum_{j=1}^{n} q_{ij} x_i x_j + \sum_{i=1}^{n} c_i x_i , \tag{3.5}
$$

and all the functions g_i (with $i = 1, \ldots, m$) are linear (see, for instance, Exercise 113). Then, Problem (3.2) is called Quadratic Programming (QP) problem. Generally, without further assumptions on the structure of the matrix Q, a QP problem has a computational effort slightly greater than that of a Linear Programming problem.

Using the notations provided by the MATLAB Help, one can express a QP problem as follows

$$
\left\{
\begin{array}{ll}
\min & \frac{1}{2} x^T H x + f^T x \\
\text{s.t.} & \\
& Ax \leq b \\
& A_{eq} x = b_{eq} \\
& l_b \leq x \leq u_b ,
\end{array}
\right. \tag{3.6}
$$

where H is an $n \times n$ matrix and represents the quadratic part of the objective function. The vectors f, b, b_{eq}, x, l_b, and u_b, and the matrices A and A_{eq} have the same task as in Problem (3.4). This class of problems is very useful in financial applications. For instance, the well-known Mean-Variance portfolio selection model is a QP problem, as shown in Chapter 4.

To solve a QP problem, the Optimization toolbox of MATLAB provides the built-in function `quadprog`, which has the following syntax:

```
[x,fval] = quadprog(H,f,A,b,Aeq,beq,l_b,u_b) ,
```

where H must be a symmetric matrix.

If the matrix H is not symmetric, then the same quadratic form $x^T H x$ can be obtained by a symmetric matrix $\overline{H}$ associated to H, called symmetric part of matrix H, where

$$
\overline{H} = \frac{1}{2} \left(H + H^T \right) .
$$

Indeed, H and $\overline{H}$ determine the same quadratic form:

$$
\begin{aligned}
x^T \overline{H} x &= \frac{1}{2} x^T \left(H + H^T \right) x \\
&= \frac{1}{2} \left(x^T H x + x^T H^T x \right) \\
&= \frac{1}{2} \left(x^T H x + x^T H x \right) \\
&= x^T H x
\end{aligned}
$$

For the other inputs, the same considerations hold as for the built-in function `linprog`. Furthermore, the outputs x and fval represent the optimal solution x^* and the optimal value $\frac{1}{2} x^{*T} H x^* + f^T x^*$ of Problem (3.6), respectively. For more details, see the MATLAB Help.

Exercise 113 *Solve the following QP problem, in order to find the optimal solutions x_1^* and x_2^* that minimize the quadratic objective function.*

$$\begin{cases} \min & x_1^2 + x_2^2 - x_1 x_2 - 2x_1 - 6x_2 \\ \text{s.t.} & \\ & -x_1 - x_2 \geq -2 \\ & -x_1 + 2x_2 \leq 2 \\ & 2x_1 + x_2 \leq 3 \\ & x_1, x_2 \geq 0 \end{cases}$$

Sol.: $[x^* = (0.6667, 1.3333) \, ; f(x^*) = -8]$. See script S_Quadprog.

3.2.3   Non-Linear Programming

As mentioned above, in the general case, when at least one of the functions f or g_i (with $i = 1, \ldots, m$) is not linear, Problem (3.2) is called Non-Linear Programming (NLP) problem.

Remark 114 (LP vs. NLP)

- *In NLP problems, one may find points of local minimum or maximum that can not be global. In PL problems, if a point is a local minimum or maximum, then it is also global.*

- *The feasible region of an NLP problem can be any shape (not necessarily a polyhedron as in LP), and the optimal solution can be anywhere (not necessarily on a vertex of a polyhedron as in LP).*

- *The algorithms used to solve NLP problems are generally much less efficient than those for LP problems. Furthermore, as previously mentioned, the solutions are not always global, but they could be only stationary points, more precisely Karush-Kuhn-Tucker (KKT) points (see, e.g., Cornuejols and Tütüncü, 2006).*

To solve an NLP problem, the Optimization toolbox of MATLAB provides several built-in functions. Here, we only present the function fmincon, which attempts to find a minimum of a multivariable scalar function on a feasible set determined by nonlinear constraints.

Using the notations provided by the MATLAB Help, one can express an NLP problem as follows

$$\begin{cases} \min & f(x) \\ \text{s.t.} & \\ & c(x) \leq 0 \\ & c_{eq}(x) = 0 \\ & Ax \leq b \\ & A_{eq}x = b_{eq} \\ & l_b \leq x \leq u_b \end{cases} \qquad (3.7)$$

where $c(x)$, $c_{eq}(x)$ and $f(x)$ are any possible kind of scalar functions. To solve a constrained NLP problem, the Optimization toolbox provides the built-in function fmincon that has the following syntax

```
[x,fval] = fmincon(fun,x0,A,b,Aeq,beq,l_b,u_b,nonlcon) ,
```

where fun and nonlcon are respectively the function to be minimized, and the functions that determine the nonlinear inequality constraints $c(x) \leq 0$ and the nonlinear equality constraints $c_{eq}(x) = 0$. Both fun and nonlcon can be represented by an anonymous function (see Section 1.1.6) or a generic MATLAB function (see Section 1.2). The input x0 is the starting point for the algorithm, while for the inputs A, b, Aeq, beq, l_b, u_b, the same considerations hold as for the built-in functions linprog and quadprog. Furthermore, the outputs x and fval represent the optimal solution x^* and the optimal value $f(x^*)$ of Problem (3.7), respectively. For more details, see the MATLAB Help.

Additionally, in the MATLAB functions linprog, quadprog and fmincon, one can include a set of options as input (e.g., the type of optimization algorithm, the maximum number of iterations, tolerance levels, etc.) using the built-in functions optimoptions and optimset. For more details, see the MATLAB Help.

Exercise 115 *Solve the following problem using* x01 = [2,4,8] *as starting point, then try with* x02 = [-8,0,-3].

$$\begin{cases} \min & f(x) = x_1 * x_2 * x_3 \\ \text{s.t.} & \\ & x_1 + 3x_2 + 4x_3 \leq 0 \\ & -2x_1 - x_2 - 0.5x_3 \leq 50 \end{cases}$$

Sol.: using x01 = [2,4,8] $[x^* = (-28.38, 5.15, 3.24)$, $f(x^*) = -472.63]$; using x02 = [-8,0,-3] $[x^* = (-8.33, -16.67, -33.33)$, $f(x^*) = -4629.63]$. See Script S_NLP and Function F_Nonlin.

105

3.3 Multi-objective optimization

Multi-objective optimization consists in solving problems having two or more different objectives. Such problems aim at simultaneously optimizing possibly conflicting goals by means of an appropriate trade-off between the two or more objective functions considered. For instance, in finance a relevant problem is to construct a portfolio of assets while satisfying two conflicting goals, the return maximization and the risk minimization (see Chapter 4).

One can formally express a generic bi-objective problem in the following equivalent manners:

$$\begin{cases} \min & f_1(x) \\ \max & f_2(x) \\ \text{s.t.} \\ & x \in C \end{cases} \Leftrightarrow \begin{cases} \min & (f_1(x), -f_2(x)) \\ \text{s.t.} \\ & x \in C \end{cases} \tag{3.8}$$

where C is the feasible set, that can be identified by a system of equalities/inequalities that represents the constraints of the problem, just like in LP or NLP problems. Multi-objective optimization problems can be graphically represented through two spaces:

1. the *decisional variable* space by which it is possible to identify the feasible set C of the problem;

2. the space of the *objective functions values* computed for each feasible solution $x \in C$.

The latter item implies that, for each feasible point x of the *decisional variable* space, we can obtain a corresponding point in the space of the *objective functions values*. For instance, considering Problem (3.8), for each $x \in C$ we can plot the corresponding point in the bi-dimensional space determined by the objective functions $f_1(x)$ and $f_2(x)$.

Note that, in an optimization problem with a single-objective function, the space of the *objective functions values* is represented by a straight-line, and the objective function corresponding to the optimal solutions (if more than one) falls into the same value, the global optimum.

To treat multi-objective problems, the first step is to clarify what an *optimal solution* is. For this purpose, we can proceed by exclusion, using Fig. 3.1 to support intuition. Let $\tilde{x} = (\tilde{x}_1, \tilde{x}_2)$, a feasible solution in C (belonging to the *decisional variable* space), and let $\tilde{f} = (f_1(\tilde{x}_1), f_2(\tilde{x}_2))$, the corresponding point in F (belonging to the *objective functions values* space), see Fig. 3.1. If we consider a point of F belonging to the dashed area, it is surely better than $\tilde{f}$

106

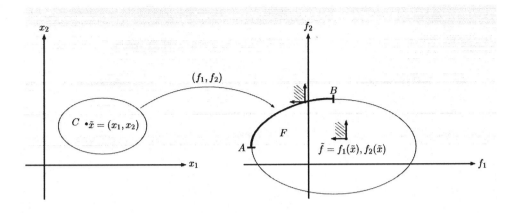

Figure 3.1: Example of decision variables space (left) and of objective function values space (right) for Problem (3.8).

both in terms of f_1 and of f_2. Then, from the viewpoint of the optimization of both objective functions, $\tilde{x}$ is not a "good" solution. As highlighted in Fig. 3.1, the feasible solutions that cannot be improved with respect to both goals belong to the bold line. Such solutions are said to be *pareto optimal* or *efficient* points of the multi-objective problem, and the set of *pareto optimal* points is called *efficient frontier*. The *efficient frontier* is a curve in the case of two objective functions, a surface with three goals, and, more generally, an iper-surface in the case of four or more objective functions.

Definition 116 (Efficient solution) *A feasible solution to a multi-objective problem is said to be efficient or Pareto optimal if there is no other feasible solution that is equally good with respect to each objective function, and better than at least one objective.*

In other words, a point is definitely inefficient if one can achieve a better value for an objective function without penalizing the others. Therefore, given the example in Fig. 3.1, a feasible solution $x \in C$ is Pareto-optimal if and only if there is not another point guaranteeing a better value for both f_1 and f_2.

A similar argument can be followed to identify the pareto-optimal points if both objective functions need to be maximized as follows

$$\begin{cases} \max & f_1(x) \\ \max & f_2(x) \\ \text{s.t.} & \\ & x \in C \end{cases} \tag{3.9}$$

For Problem (3.9), the *decisional variable* space, the *objective functions values* space and the efficient frontier are shown in Fig. 3.2.

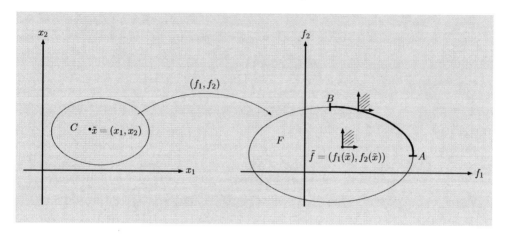

Figure 3.2: Example of decision variables space (left) and of objective function values space (right) for Problem (3.9).

Among the points belonging to the efficient frontier (all *"optimal"* in the sense described above) we can detect a point that is better than the others, by introducing a utility function that measures the importance of a goal with respect to another, assigning a weight to each objective function.

3.3.1 Efficient solutions and the efficient frontier

In this section we propose two methods for detecting the efficient frontier of a multi-objective optimization problem.

A first method is the ε-constraint method (see, e.g., Miettinen (2012)). Given Problem (3.8), we can fix a minimum value k to be achieved for the maximization objective. Therefore, we add the new constraint $f_2(x) \geq k$, thus narrowing the feasible set C, but optimizing just one objective function. The resulting optimization problem is the following

$$
\begin{cases}
\min & f_1(x) \\
s.t & \\
& f_2(x) \geq k \\
& x \in C
\end{cases}
\tag{3.10}
$$

Clearly, one can solve Problem (3.10) exploiting usual methods of single-objective optimization. Given a range of values k, one can find the corresponding efficient points. As shown in Fig. 3.3, these points may be joined to obtain a polygonal, which is a good approximation of the efficient frontier.

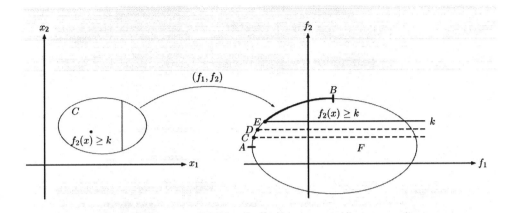

Figure 3.3: Geometrical representation of Problem (3.8) transformation by adding a constraint for the second objective function.

Here, we report a second method to solve a multi-objective problem. Consider again Problem (3.8). One can assign a *weight* $\lambda \geq 0$ to the second objective $f_2(x)$, indicating its importance with respect to the first objective $f_1(x)$ (which conventionally has weight equal to 1).
In this way, one can gather the two objectives in just one function as follows:

$$\begin{cases} \min & f_1(x) - \lambda f_2(x) \\ s.t \\ & x \in C \end{cases} \tag{3.11}$$

with $\lambda \geq 0$. The minus sign is due to the fact that $\max f_2(x) = -\min -f_2(x)$, as described at the beginning of this chapter.
From Problem (3.11), we can see that when λ is large, the predominant goal will be the maximization of $f_2(x)$; while, if λ is small, then the predominant goal in the problem will be the minimization of $f_1(x)$. All the intermediate cases can be found by varying $\lambda \in [0, +\infty)$.
Let us analyze this second method geometrically. As shown in Fig. 3.4, in the space of the *objective functions values*, $u = f_1(x) - \lambda f_2(x)$ represent the indifference curves of a (linear) utility function, where the highest values of u are obtained along the direction of the vector $(1, -\lambda)$. Therefore, in the case of minimization of u, for a fixed λ the efficient point E can be found from the first utility function (in the opposite direction of the vector $(1, -\lambda)$) that is

109

perpendicular to $(1, -\lambda)$. Clearly, E must also belong to the feasible set of the space of the *objective functions values*.

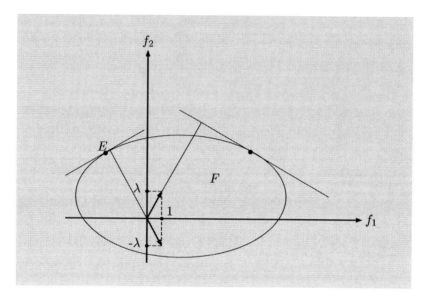

Figure 3.4: Geometrical representation of the scalarization procedure of Problem (3.8).

Chapter 4

Portfolio Optimization

The selection of an appropriate portfolio of assets is an essential component of fund management. Although a part of portfolio selection decisions is still taken on a qualitative basis, quantitative approaches to decisions under uncertainty are becoming more widely adopted. Portfolio optimization aims at choosing the weights of a given set of assets, so that the selected portfolio is the best one according to specific criteria.

The key problem in asset allocation is to select a portfolio with appropriate features in terms of gain and risk. Among academics and practitioners, different measures of gain (expected return, median return, absolute or relative wealth, etc.) and risk (variance, semivariance, Mean Absolute Deviation, Conditional Value-at-Risk, etc.) have been proposed.

From a mathematical viewpoint, the objective function $f(x)$ described in the previous chapter could represent the portfolio return or its risk; thus, $f(x)$ should be maximized or minimized, respectively. Furthermore, equality and/or inequality constraints can be added to the portfolio optimization, depending on the model, on the market conditions and so on. For instance, in all models considered we assume short selling is not allowed, namely in the corresponding optimization problem we impose the non-negativity of the portfolio weights.

In the first part of this chapter, we start dealing with prices and returns of assets. Then, we cover risk-gain analysis, which is the central part of the chapter. Starting from the seminal model of Markowitz (1952, 1959), who is regarded as the father of Modern Portfolio Theory, we later consider other popular portfolio selection models, which aim at overcoming computational and (alleged) theoretical drawbacks of the former. Finally, the last part of the chapter is dedicated to immunization. Note that portfolio optimization is applied to equities, while immunization is applied to bonds.

For more on Portfolio Selection, interested readers can consult, e.g., Castellani et al (2005b); Elton et al (2009); Meucci (2009b).

4.1 Portfolio of equities: prices and returns

Consider an investment universe of n assets in a single-period time horizon, identified by the time interval $[t, s]$. At time t one can buy or sell one or more financial contracts, for instance equities. Let us denote by $P_{k,t}$ the (known) price of a generic equity k at time t, and $P_{k,s}$ the future price, referred to the same equity k, at time s. We can represent a portfolio by means of a vector $q = [q_1, q_2, \ldots, q_n]$, where the kth element is the number of shares of the stock k, and generally $q \in \mathbb{Z}^n$. The price of a portfolio P_P, identified by the vector q, can be represented respectively, at time t and s, as follows

$$P_{P,t} = \sum_{k=1}^{n} q_k P_{k,t} \quad \text{and} \quad P_{P,s} = \sum_{k=1}^{n} q_k P_{k,s} \, .$$

However, one can also represent a portfolio by means of the fraction of capital invested in each stock. If $P_{P,t} \neq 0$, then it is possible to identify the portfolio by a vector $x = [x_1, x_2, \ldots, x_n]$, where

$$x_k = \frac{q_k P_{k,t}}{P_{P,t}} \qquad \text{with } k = 1, \ldots, n \qquad (4.1)$$

that is the percentage of capital invested in the asset k. Clearly, the elements of x must sum to one, namely $\sum_{k=1}^{n} x_k = 1$.

Often, in financial markets, investment opportunities are expressed in terms of returns, instead of prices. This is because the equity return can be considered approximately a stationary random variable (see, e.g., Meucci, 2009b, and references therein). Given the above notation, the linear return of a generic stock is

$$R_{k,s} = \frac{P_{k,s} - P_{k,t}}{P_{k,t}} = \frac{P_{k,s}}{P_{k,t}} - 1 \, . \qquad (4.2)$$

Note that $R_{k,s}$ is a random variable and, unlike prices, can assume negative values. The only limitation is that, by definition, the linear returns must be greater than or equal to -1. Indeed, for some financial models (e.g., econometrics models) the returns are defined as *log-returns*, namely $\widetilde{R}_{k,s} = \ln \frac{P_{k,s}}{P_{k,t}}$, and virtually $\widetilde{R}_{k,s} \in (-\infty, +\infty)$. However, for portfolio selection models the linear returns are preferred to the logarithmic ones, due to their mathematical tractability. Furthermore, there exists the following relationship between linear and logarithmic returns

$$\widetilde{R}_{k,s} = \ln\left(1 + R_{k,s}\right), \tag{4.3}$$

and for small values of $R_{k,s}$, it is straightforward to demonstrate that $\widetilde{R}_{k,s} \simeq R_{k,s}$. Indeed, using the Taylor expansion of a differentiable scalar function $f(x)$, we have

$$f(x) = f\left(x_0\right) + f'\left(x_0\right)\left(x - x_0\right) + \ldots$$

In the case of Expression (4.3), if $x \simeq x_0 = 0$, we have

$$\ln(1 + x) \simeq \ln(1 + 0) + \frac{1}{1 + 0}(x - 0) \Rightarrow \ln(1 + x) \simeq x$$

$$\Rightarrow \widetilde{R}_{k,s} \simeq R_{k,s}.$$

As mentioned before, the choice of linear returns for asset allocation is due to their mathematical tractability, because there exists a linear relation between the return of a portfolio and the return of its selected stocks. In fact, given $P_{P,t} = \sum_{k=1}^{n} q_k P_{k,t}$, $P_{P,s} = \sum_{k=1}^{n} q_k P_{k,s}$, and $P_{k,s} = P_{k,t}\left(1 + R_{k,s}\right)$, we can then write

$$R_{P,s} = \frac{P_{P,s} - P_{P,t}}{P_{P,t}} = \frac{P_{P,s}}{P_{P,t}} - 1.$$

Thus,

$$R_{P,s} = \frac{\sum_{k=1}^{n} q_k P_{k,t}\left(1 + R_{k,s}\right)}{\sum_{k=1}^{n} q_k P_{k,t}} - 1 = \frac{\sum_{k=1}^{n} q_k P_{k,t} R_{k,s}}{\sum_{k=1}^{n} q_k P_{k,t}} = \sum_{k=1}^{n} x_k R_{k,s},$$

where $x_k = \frac{q_k P_{k,t}}{\sum_{k=1}^{n} q_k P_{k,t}}$. While, in the case of log-returns, we have

$$\widetilde{R}_{P,s} = \ln\left(\frac{P_{P,s}}{P_{P,t}}\right) = \ln\left(\frac{\sum_{k=1}^{n} q_k P_{k,s}}{\sum_{k=1}^{n} q_k P_{k,t}}\right) \neq \sum_{k=1}^{n} x_k \ln\left(\frac{P_{k,s}}{P_{k,t}}\right).$$

Exercise 117 (Prices and returns) *Let P be a $(1 \times T)$ vector of prices for a certain asset, then calculate the time series of linear returns $R_t = \frac{P_t - P_{t-1}}{P_{t-1}}$ and that of the logarithmic returns $\widetilde{R}_t = \ln\left(\frac{P_t}{P_{t-1}}\right) = \ln\left(P_t\right) - \ln\left(P_{t-1}\right)$, for $t = 2, ..., T$. Hint: see the built-in function* diff. *Finally, try to calculate these returns using a* for *loop.*
Sol.: See Script S_Prices_Returns.

Now, we deal with the portfolio expected return μ_P and its variance σ_P^2, which will widely use as gain and risk measures, respectively, for the portfolio selection approach based on the Risk-Return analysis (see Section 4.2).

As mentioned above, the portfolio return is a random variable. More precisely, the (linear) portfolio return is the sum of n random variables, which are the available stock returns in the market, weighted by their respective shares (in percentage), $R_P = \sum_{k=1}^{n} x_k R_k$. Then, since the expected value is a linear operator (see Remark 85) the portfolio expected return μ_P can be expressed as follows

$$
\begin{aligned}
\mu_P &= E\left(R_P\right) = E\left(\sum_{k=1}^{n} x_k R_k\right) \\
&= \sum_{k=1}^{n} x_k E\left(R_k\right) = \sum_{k=1}^{n} x_k \mu_k,
\end{aligned} \tag{4.4}
$$

where, therefore, μ_k indicates the expected return of the asset k.

Before defining the variance of a portfolio, it is worth to recall the definition of the covariance between two generic assets' returns R_k and R_j

$$
\sigma_{kj} = cov(R_k, R_j) = E\left[(R_k - \mu_k)(R_j - \mu_j)\right]. \tag{4.5}
$$

Consequently, the variance of the return of a stock is $\sigma_{kk} = \sigma_k^2 = Var[R_k]$.

Remark 118 (Covariance) *The covariance between two generic random variables X and Y is*

$$
\begin{aligned}
\sigma_{XY} &= cov(X, Y) \\
&= E[(X - E[X])(Y - E[Y])] \\
&= E[XY] - E[X]E[Y].
\end{aligned}
$$

For discrete random variables with equally likely realization values, we have $\sigma_{XY} = \frac{1}{n}\sum_{k=1}^{n}(x_k - \mu_X)(y_k - \mu_Y)$, while for continuous r.v. $\sigma_{XY} = \int_{-\infty}^{+\infty}(x - \mu_X)(y - \mu_Y)f_{X,Y}(x,y)dxdy$, where $f_{X,Y}(x,y)$ is the joint probability density function of the bivariate r.v. (X,Y).

From the linearity of the expectation operator and from Expression (4.5), the variance of a portfolio with n stocks can be expressed as follows

$$
\begin{aligned}
\sigma_P^2 &= Var\left[R_P\right] \\
&= E\left[(R_P - \mu_P)^2\right] = E\left[\left(\sum_{k=1}^{n} x_k(R_k - \mu_k)\right)^2\right] \\
&= E\left[\left(\sum_{k=1}^{n} x_k(R_k - \mu_k)\right)\left(\sum_{j=1}^{n} x_j(R_j - \mu_j)\right)\right] \\
&= E\left[\sum_{k=1}^{n}\sum_{j=1}^{n} x_k x_j (R_k - \mu_k)(R_j - \mu_j)\right] \\
&= \sum_{k=1}^{n}\sum_{j=1}^{n} x_k x_j E\left[(R_k - \mu_k)(R_j - \mu_j)\right] \\
&= \sum_{k=1}^{n}\sum_{j=1}^{n} x_k x_j \sigma_{kj} \, . \tag{4.6}
\end{aligned}
$$

In matrix form we have

$$
\sigma_P^2 = Var\left(R_P\right) = x^T \Sigma x \, , \tag{4.7}
$$

where Σ is the covariance matrix, namely

$$
\Sigma = \begin{pmatrix}
\sigma_{11} & \sigma_{12} & \cdots & \sigma_{1n} \\
\sigma_{21} & \sigma_{22} & \cdots & \sigma_{2n} \\
\vdots & & & \\
\sigma_{n1} & \sigma_{n2} & \cdots & \sigma_{nn}
\end{pmatrix} . \tag{4.8}
$$

Note that since $\sigma_{kj} = \sigma_{jk}$, Σ is symmetric.

Exercise 119 (Import financial data) *Solve the following points.*

1. *From the* Yahoo!Finance *website, download the historical prices of the following assets:* ENEL.MI *(Enel),* ENI.MI *(Eni),* FNC.MI *(Finmeccanica) and* G.MI *(Generali). Import the data into the workspace and calculate their returns using the adjusted prices. Generate a matrix* Ret *of the assets' returns, and save it in a .xls file, named* Returns.xls. *The length of time series and their frequency (monthly, weekly, daily) is arbitrary. Hint: in the MATLAB Help see the built-in functions* importdata *and* flipud.

> *2. For each asset, plot the linear return as a function of time, and check if there are possible bugs in the data.*
>
> *3. Using the matrix* `Ret`, *compute the mean, standard deviation, skewness, and kurtosis of the assets returns. Furthermore, calculate the covariance matrix of these returns.*
>
> *Finally, save the workspace in a .mat file, named*
> `Import_Yahoo_Workspace.mat`.
>
> **Sol.**: See Script `S_Import_Yahoo`.

4.2 Risk-return analysis

4.2.1 Elements of Expected Utility Theory

In Expected Utility Theory an investor evaluates a random portfolio wealth W_P with respect to her preferences, represented by a utility function u (see, e.g., Levy, 2015; Ingersoll, 1987). She will choose the portfolio with the maximum expected utility. Formally, we have

$$\max_{x \in C} E[u(W_P(x))], \qquad (4.9)$$

where x is the vector of portfolio weights and C is a set of feasible portfolios. In this book we typically consider the long-only portfolios satisfying the budget constraint, namely $x \in \Delta = \{x \in \mathbb{R}^n : \sum_{i=1}^{n} x_i = 1, x_i \geq 0, \quad i = 1, \dots, n\}$. Portfolio selection is a typical decision problem under uncertainty, where the drivers of uncertainty are the asset returns. In general, the choices under conditions of uncertainty depends on several factors and criteria. The expected utility maximization allows for a reduction in the complexity of the problem of choice since it consists in a global optimization problem with a single objective. In a nutshell, the maximization of the expected utility provides a synthesis (w.r.t. the preference of an investor) of all the criteria of choice. However, to better understand the structure of this decision problem, it could be interesting to disaggregate the single objective in several partial objectives. This approach naturally leads to highlight the close links between the expected utility criterium and the Risk-Gain analysis as shown in the following. Similar to Allais (1984), we can introduce the following generalized risk measure

$$\mathcal{R}(W_P(x)) = u(E[W_P(x)]) - E[u(W_P(x))] \qquad (4.10)$$

which, by *Jensen's inequality* (Jensen, 1906), is not negative for a risk-averse investor (i.e., an investor with a utility function u such that $u' > 0$ and $u'' < 0$), and it is null only when the portfolio wealth is nonrandom or when the utility function u is linear. Then, the global objective of maximizing expected utility can be disaggregated in two partial objectives: the maximization of $u(E[W_P(x)])$ and the minimization of $\mathcal{R}(W_P(x))$ as follows.

$$\max_{x \in \Delta} E[u(W_P(x))] = \max_{x \in \Delta} u(E[W_P(x)]) - \mathcal{R}(W_P(x)) \qquad (4.11)$$

Therefore, Problem (4.11) is equivalent to the following bi-objective optimization problem

$$\begin{aligned} &\min && \mathcal{R}(W_P(x)) \\ &\max && u(E[W_P(x)]) \\ &s.t. && x \in \Delta \end{aligned} \qquad (4.12)$$

which could be interpreted and solved by means of the standard Risk-Gain analysis in the $\mathcal{R} - UE$ (Risk-Gain) plane. Note that, since $u(\cdot)$ is a monotonic increasing function with respect to its argument, we can directly maximize $E[W_P(x)]$. Hence, Problem (4.12) can be restated as

$$\begin{aligned} &\min && \mathcal{R}(W_P(x)) \\ &\max && E[W_P(x)] \\ &s.t. && x \in \Delta \end{aligned} \qquad (4.13)$$

However, since the utility function could vary among individuals, the choice is valid only for investors whose utility function takes on that specific form. In order to overcome this subjectivity, some scholars have replaced $\mathcal{R}$ with different risk measures, showing that this choice (in particular the portfolio variance) can be still a good approximation of the EU criterion (Carleo et al, 2017; Markowitz, 2012; Markowitz and Blay, 2013; Markowitz, 2014; Meucci, 2009b). In fact, in portfolio selection, typically, one can distinguish a first step of the Risk-Gain analysis (see, e.g., Markowitz, 1959; Elton et al, 2009; Blay and Markowitz, 2013) where the efficient portfolios, namely the Pareto optimal solutions, are detected. Among the Pareto optimal portfolios, one can adopt, with respect to risk or gain, preference criteria that, possibly, are introduced in the second step of the gain-risk analysis.

As we will see in the following sections, we can identify the portfolio efficient frontier by solving the bi-objective optimization problem (4.13), where the generalized risk measure $\mathcal{R}$ is replaced by an nonsubjective risk measure, such as variance (see Section 4.2.3), mean absolute deviation (see Section 4.2.5), max loss (see Section 4.2.6), Conditional Value-at-Risk (see Section 4.2.8), Gini's mean difference (see Section 4.2.9). Thus, we consider the Risk-Gain analysis as the objective phase of the portfolio selection procedure, because it depends solely on the characteristics of the market random returns. Hence, this part of the analysis can be regarded the same for any rational operator.

4.2.2 General Framework

As stated earlier, the start of Modern Portfolio Theory generally coincides with the publication of the Mean-Variance (MV) model of Markowitz (1952, 1959). The purpose of Modern Portfolio Theory is to study the optimal capital allocation among stocks available on the market, by focusing on the trade-off between risk and return. However, since the asset return is generally a random variable, a choice criterion under uncertainty conditions is needed.

Let us discuss the concept of Risk-Return approach from another point of view w.r.t. the previous section, by means of an example.

> **Example 120** *Suppose that an operator can choose between two investments described by the following random variables: $X = \{50€, 150€\}$ with probabilities of occurrence $p_X = \{\frac{1}{2}, \frac{1}{2}\}$, and $Y = \{0€, 200€\}$ with $p_Y = \{\frac{1}{2}, \frac{1}{2}\}$. It is clear that both the random variables have the same expected value $E[X] = E[Y] = 100$, but it is intuitive to consider the second investment riskier than the first. Indeed, the dispersion of Y around its expected value is larger than that of X. A (statistical) synthetic index that can represent the dispersion is the variance (see Section 2.5), and it can be used as a possible risk measure related to the uncertainty. Hence, the greater the variance, the larger risk.*

Given the expected value and the variance of a r.v. that represents an investment, the goal of a rational investor will be to maximize the return while minimizing the risk. This is the strategy on which the mean-variance criterion is based. Mathematically, the bet $X = \{x_t; p_t\}_{t=1,...T}$ is preferable to the bet $Y = \{y_t, p_t\}_{t=1,...T}$ if:

$$\begin{cases} \sigma_X^2 \leq \sigma_Y^2 \\ \mu_X \geq \mu_Y \end{cases}$$

where at least one of the inequalities is strict.

Now let us consider the Mean-Variance plane and a generic investment represented by a point P on this plane. As shown in Fig. 4.1, we can distinguish three areas of preferences:

1. an area where all investments are preferred to the bet P;

2. an area where all investments are not preferred to P;

3. an area in which it is not possible to establish an order of preference by means of the Mean-Variance approach.

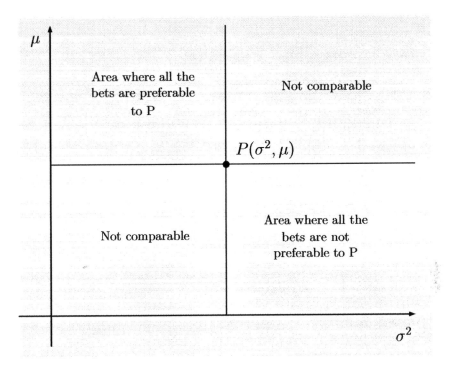

Figure 4.1: Preferences in the Mean-Variance plane.

In the next section, we will describe the Mean-Variance model, starting from the general properties of variance and volatility. We will introduce the Markowitz's approach in the case of two risky securities, and then we will extend it to more general cases.

4.2.3 Mean-Variance model

The classical Mean-Variance (MV) portfolio optimization model aims at determining the fraction x_i of a given capital to be invested in each asset i belonging to a given market, in order to minimize the risk of the return of the whole portfolio, identified with its variance, while restricting the expected return of the portfolio to attain at least a specified value.

Before describing the Mean-Variance approach in the simple case of two risky assets, we present some general properties of the portfolio variance and volatility.

Remark 121 (Variance and volatility properties)
As shown in Section 4.1, the portfolio variance $\sigma_P^2(x) = x^T \Sigma x$ (where Σ is the covariance matrix), and is a risk measure $\sigma_P^2 : \mathbb{R}^n \to \mathbb{R}_+^0$ that satisfies the following properties:

119

- $\sigma_P^2\left(\lambda x\right) = \lambda^2\sigma_P^2\left(x\right) \qquad \forall\, x \in \mathbb{R}^n, \forall\, \lambda \in \mathbb{R}$ (homogeneity of degree 2)

- $\sigma_P^2\left(\lambda x + (1-\lambda)y\right) \leq \lambda\sigma_P^2\left(x\right) + (1-\lambda)\sigma_P^2\left(y\right) \quad \forall\, x, y \in \mathbb{R}^n,\ \forall\, \lambda \in [0,1]$ (convexity)

Proof. *For the first property we have* $\sigma_P^2\left(\lambda x\right) = \lambda x^T \Sigma \lambda x = \lambda^2\sigma_P^2\left(x\right)$. *To demonstrate the convexity of variance, we can show that*

$$\lambda\sigma_P^2\left(x\right) + (1-\lambda)\sigma_P^2\left(y\right) - \sigma_P^2\left(\lambda x + (1-\lambda)y\right) \geq 0\,.$$

Thus, we have

$$\lambda x^T\Sigma x + (1-\lambda)y^T\Sigma y - (\lambda x^T + (1-\lambda)y^T)\Sigma(\lambda x + (1-\lambda)y) \geq 0$$
$$\Rightarrow\ \lambda x^T\Sigma x + (1-\lambda)y^T\Sigma y - \lambda^2 x^T\Sigma x - 2(1-\lambda)\lambda y^T\Sigma x - (1-\lambda)^2 y^T\Sigma y \geq 0$$
$$\Rightarrow\ \lambda(1-\lambda)x^T\Sigma x + (1-\lambda)\lambda y^T\Sigma y - 2(1-\lambda)\lambda y^T\Sigma x \geq 0$$
$$\Rightarrow\ \lambda(1-\lambda)\left[x^T\Sigma x + y^T\Sigma y - 2y^T\Sigma x\right] \geq 0$$
$$\Rightarrow\ \lambda(1-\lambda)\left[(x-y)^T\Sigma(x-y)\right] \geq 0$$

■

The portfolio volatility $\sigma_P\left(x\right) = \sqrt{\sigma_P^2\left(x\right)} = \sqrt{x^T\Sigma x}$, *and is a risk measure* $\sigma_P : \mathbb{R}^n \to \mathbb{R}_+^0$ *that satisfies the following properties:*

- $\sigma_P\left(\lambda x\right) = \lambda\sigma_P\left(x\right) \quad \forall\, x \in \mathbb{R}^n, \forall\, \lambda \geq 0$ (positive homogeneity)

- $\sigma_P\left(x + y\right) \leq \sigma_P\left(x\right) + \sigma_P\left(y\right) \quad \forall\, x, y \in \mathbb{R}^n$ (subadditivity)

However, it is worth noting that these two properties imply that the portfolio volatility is also a convex risk measure

$$\sigma_P\left(\lambda x + (1-\lambda)y\right) \leq \lambda\sigma_P\left(x\right) + (1-\lambda)\sigma_P\left(y\right) \qquad \forall\, x, y \in \mathbb{R}^n,\ \forall\, \lambda \in [0,1]\,.$$

Two risky assets

As a first step, let us assume that only two risky assets, A and B, are available in the market, and let us identify the assets A and B in the $\{\sigma^2, \mu\}$ plane by the points $\{\sigma_A^2, \mu_A\}$ and $\{\sigma_B^2, \mu_B\}$, respectively. Furthermore, let us assume that $\mu_A < \mu_B$ and $\sigma_A^2 < \sigma_B^2$, otherwise one point is preferred to the other as shown in Fig. 4.1. Therefore, as described in Expressions (4.4) and (4.6), the portfolio expected return μ_P and its variance σ_P^2 can be expressed as follows

$$\begin{aligned} \mu_P &= \mu^T x = x_A\mu_A + x_B\mu_B \\ \sigma_P^2 &= x^T\Sigma x = x_A^2\sigma_A^2 + x_B^2\sigma_B^2 + 2x_A x_B\sigma_{AB}\,, \end{aligned}$$

where $x = \begin{bmatrix} x_A \\ x_B \end{bmatrix}$ is the vector of weights, $\mu = \begin{bmatrix} \mu_A \\ \mu_B \end{bmatrix}$ is the vector of the

expected returns, and $\Sigma = \begin{bmatrix} \sigma_A^2 & \sigma_{AB} \\ \sigma_{AB} & \sigma_B^2 \end{bmatrix}$ is the variance-covariance matrix.

Furthermore, if we denote by $\rho_{AB} = \dfrac{\sigma_{AB}}{\sigma_A \sigma_B}$ the linear (Pearson) correlation between the returns of the asset A and those of the asset B, then we can rewrite the portfolio variance as

$$\sigma_P^2 = x_A^2 \sigma_A^2 + x_B^2 \sigma_B^2 + 2 x_A x_B \rho_{AB} \sigma_A \sigma_B .$$

Exercise 122 (Portfolio of 2 risky assets)
In the Script S_2assetVariance, *solve the following points.*

1. *Import the data contained into* stoxx50_0713.xls, *creating a matrix* A. *In this sheet there are the historical prices of the stocks components of the EUROSTOXX50 index. Then, define a new matrix* B *containing only the prices of the first two assets.*

2. *Compute the historical linear returns creating a matrix* ret_B.

3. *Compute the covariance matrix corresponding to the returns matrix* ret_B.

4. *Compute the variance of the portfolio composed of 30% of the first asset and 70% of the second asset.*

5. *Detect the portfolio with the minimum variance when short sales are not allowed (see the built-in function* min*). Consider also the usual budget constraint. Hint: define the vector of the portfolio weights* $x = [w, 1 - w]$ *with the weight w of the first asset and $1 - w$ of the second; therefore, if w ranges from 0 to 1 (for instance with a step of 0.05), it is possible to compute the variance $\sigma^2(w)$ of all the feasible portfolios as a function of w.*

6. *Finally, plot σ^2 vs. w and save the graph as* 2assetVariance.tif *verifying that the portfolio minimum variance corresponds to the one represented in the figure.*

Sol.: See Script S_2assetVariance.

Remark 123 (Correlation between two risky assets)
The linear (Pearson) correlation coefficient is one of the most used synthetic indices to measure the level of co-movement among pairs of random variables (or of their observed outcomes). Its values range between -1 (perfect negative correlation) and $+1$ (perfect positive correlation) passing through 0 (uncorrelated random variables). An example of correlation analysis between the returns of two assets, A and B, is shown in Fig. 4.2. At the top, the evolution over time of the returns, r_A (red line) and r_B (blue line), is exhibited for different levels of correlation: when $\rho_{AB} = -0.92$ (high anticorrelation), r_A and r_B tend to move in opposite directions; when $\rho_{AB} = 0.01$ (low correlation), r_A and r_B tend to have no relation in the upward and downward movements; when $\rho_{AB} = 0.97$ (high correlation), r_A and r_B tend to move in the same directions. At the bottom of Fig. 4.2, we report the corresponding scatter plots of r_A vs. r_B, that clearly show the level of linear correlation between the random variables analyzed. Again, if ρ_{AB} is highly negative, when r_A decreases, r_B increases, and vice versa. While, if ρ_{AB} is highly positive, when r_A increases, r_B follows. Finally, when r_A and r_B are linearly uncorrelated, the scatter plot appears as a cloud of points, which highlights the absence of co-movement between the assets returns.

Now, let us examine more in detail the set of feasible points in the Mean-Variance plane in the case of a portfolio of two risky assets. Then, we have to investigate how (σ_P^2, μ_P) vary in correspondence of the weights vector $x = (x_A, x_B)$

$$\begin{cases} \sigma_P^2 = x_A^2\sigma_A^2 + x_B^2\sigma_B^2 + 2\rho_{AB}\sigma_A\sigma_B x_A x_B \\ \mu_P = x_A\mu_A + x_B\mu_B \end{cases} \tag{4.14}$$

with the budget constraint $x_A + x_B = 1$. Furthermore, if we assume that no shortselling is allowed, then we have to impose $x_A, x_B \geq 0$, namely $x_A \in [0, 1]$ (or, equivalently $x_B \in [0, 1]$). Since we have $x_A = 1 - x_B$, Problem (4.14) depends on only one weight. Thus, the feasible region is described by the following system of equations

$$\begin{cases} \sigma_P^2 = (1 - x_B)^2\sigma_A^2 + x_B^2\sigma_B^2 + 2\rho_{AB}\sigma_A\sigma_B(1 - x_B)x_B \\ \mu_P = (1 - x_B)\mu_A + x_B\mu_B \end{cases}$$

and we can study the frontier of the portfolio by varying x_B between 0 and 1. At first, we can individuate the point of the frontier corresponding to $x_B = 0$ and $x_B = 1$:

$$\begin{cases} \sigma_P^2 = \sigma_A^2 \\ \mu_P = \mu_A \end{cases} \text{for } x_B = 0 \quad \text{and} \quad \begin{cases} \sigma_P^2 = \sigma_B^2 \\ \mu_P = \mu_B \end{cases} \text{for } x_B = 1.$$

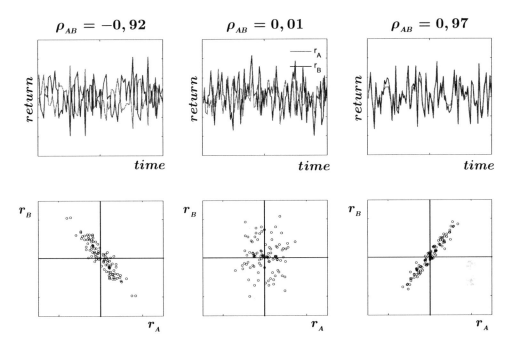

Figure 4.2: Example of different levels of correlation between two assets

Therefore, as shown in Fig. 4.3, the points $A = (\sigma_A, \mu_A)$ and $B = (\sigma_B, \mu_B)$ are the extreme points of the feasible portfolios frontier. Now, to explicitly identify the frontier, it is necessary to make assumptions on the values of the correlation coefficient.

The case of perfect positive correlation

In the case of perfectly positively correlated securities ($\rho_{AB} = 1$), the portfolio variance becomes

$$
\begin{aligned}
\sigma_P^2 &= (1 - x_B)^2 \sigma_A^2 + x_B^2 \sigma_B^2 + 2\sigma_A \sigma_B (1 - x_B) x_B \\
&= ((1 - x_B)\sigma_A + x_B \sigma_B)^2 \\
\Rightarrow \quad \sigma_P &= \sigma_A + x_B(\sigma_B - \sigma_A) \\
\Rightarrow \quad x_B &= \frac{\sigma_P - \sigma_A}{\sigma_B - \sigma_A}.
\end{aligned}
$$

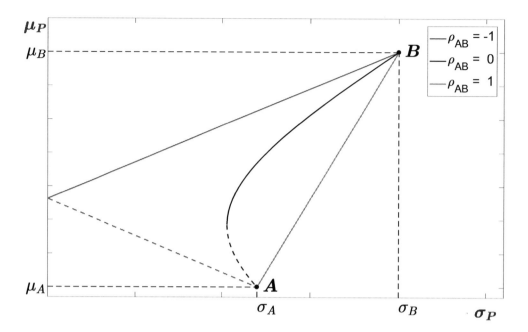

Figure 4.3: Example of efficient frontiers of a portfolio with two risky assets for different levels of correlation.

Thus, using the expression of x_B to calculate μ_P, we obtain

$$\mu_P = (1 - x_B)\mu_A + x_B\mu_B$$

$$= \mu_A + x_B(\mu_B - \mu_A)$$

$$= \mu_A + \frac{\sigma_P - \sigma_A}{\sigma_B - \sigma_A}(\mu_B - \mu_A)$$

$$\Rightarrow \mu_P = \underbrace{\frac{(\mu_A\sigma_B - \mu_B\sigma_A)}{\sigma_B - \sigma_A}}_{\text{intercept}} + \sigma_P \underbrace{\frac{\mu_B - \mu_A}{\sigma_B - \sigma_A}}_{\text{slope}}.$$

Therefore, the frontier (that for $\rho_{AB} = 1$ is also *efficient*) is represented by the straight solid line with slope $\theta = \dfrac{\mu_B - \mu_A}{\sigma_B - \sigma_A}$ joining the points $A = (\sigma_A, \mu_A)$ and $B = (\sigma_B, \mu_B)$ (red line in Fig. 4.3).

The case of perfect negative correlation

In the case of perfectly negatively correlated securities ($\rho_{AB} = -1$), we have

$$\sigma_P^2 = (1 - x_B)^2 \sigma_A^2 + x_B^2 \sigma_B^2 - 2\sigma_A \sigma_B (1 - x_B) x_B$$

$$= ((1 - x_B)\sigma_A - x_B \sigma_B)^2$$

$$\Rightarrow \sigma_P = |\sigma_A - x_B(\sigma_A + \sigma_B)|$$

$$= \begin{cases} \sigma_A - x_B(\sigma_A + \sigma_B) & \text{if } 0 \leq x_B \leq \dfrac{\sigma_A}{\sigma_A + \sigma_B} \\[3mm] -\sigma_A + x_B(\sigma_A + \sigma_B) & \text{if } \dfrac{\sigma_A}{\sigma_A + \sigma_B} < x_B \leq 1 \end{cases}$$

$$\Rightarrow x_B = \begin{cases} \dfrac{\sigma_A - \sigma_P}{\sigma_A + \sigma_B} & \text{if } 0 \leq x_B \leq \dfrac{\sigma_A}{\sigma_A + \sigma_B} \\[3mm] \dfrac{\sigma_A + \sigma_P}{\sigma_A + \sigma_B} & \text{if } \dfrac{\sigma_A}{\sigma_A + \sigma_B} < x_B \leq 1 \end{cases}$$

Thus, using the expression of x_B to calculate μ_P, we have

$$\begin{cases} \mu_P = \dfrac{\mu_A \sigma_B + \mu_B \sigma_A}{\sigma_B + \sigma_A} - \dfrac{\mu_B - \mu_A}{\sigma_B + \sigma_A}\sigma_P & \text{if } 0 \leq x_B \leq \dfrac{\sigma_A}{\sigma_A + \sigma_B} \\[3mm] \mu_P = \dfrac{\mu_A \sigma_B + \mu_B \sigma_A}{\sigma_B + \sigma_A} + \dfrac{\mu_B - \mu_A}{\sigma_B + \sigma_A}\sigma_P & \text{if } \dfrac{\sigma_A}{\sigma_A + \sigma_B} < x_B \leq 1 \end{cases}$$

Therefore, the frontier is a piecewise linear curve, where the straight solid line with positive slope represents the *efficient* frontier (solid blue line in Fig. 4.3). It is interesting to observe that, in the case of perfect negative correlation, it is possible to construct a portfolio with zero risk, namely with $\sigma_P = 0$, and with the expected return $\mu_P = \dfrac{\mu_A \sigma_B + \mu_B \sigma_A}{\sigma_B + \sigma_A}$.

The case of zero correlation

Finally, in case of two assets with zero correlation ($\rho_{AB} = 0$), the expected portfolio return is

$$\mu_P = (1 - x_B)\mu_A + x_B\mu_B$$

$$= \mu_A + x_B(\mu_B - \mu_A)$$

$$\Rightarrow x_B = \frac{\mu_P - \mu_A}{\mu_B - \mu_A} \; ; \qquad 1 - x_B = \frac{\mu_B - \mu_P}{\mu_B - \mu_A}$$

while for the portfolio variance we have

$$
\sigma_P^2 = (1 - x_B)^2 \sigma_A^2 + x_B^2 \sigma_B^2
$$

$$
= \left(\frac{\mu_B - \mu_P}{\mu_B - \mu_A} \right)^2 \sigma_A^2 + \left(\frac{\mu_P - \mu_A}{\mu_B - \mu_A} \right)^2 \sigma_B^2
$$

$$
\Rightarrow (\mu_B - \mu_A)^2 \sigma_P^2 = (\mu_B - \mu_P)^2 \sigma_A^2 + (\mu_P - \mu_A)^2 \sigma_B^2 .
$$

After some algebraic manipulation, we obtain the following equation of the frontier of the feasible portfolios set when the two risky assets are uncorrelated

$$
\sigma_P^2 = \frac{\sigma_A^2 + \sigma_B^2}{(\mu_B - \mu_A)^2} \mu_P^2 - 2 \frac{\mu_B \sigma_A^2 + \mu_A \sigma_B^2}{(\mu_B - \mu_A)^2} \mu_P + \frac{\mu_B^2 \sigma_A^2 + \mu_A^2 \sigma_B^2}{(\mu_B - \mu_A)^2} . \tag{4.15}
$$

In the Mean-Variance plane Eq. (4.15) is a horizontal parabola, while in the Mean-Volatility plane Eq. (4.15) is represented by an arc of hyperbole (black line in Fig. 4.3). Furthermore, note that the *efficient* portfolios are those belonging to the solid black line.

Remark 124 *If shortsellings are allowed, then the frontier will extend beyond the extreme points $A = (\sigma_A^2, \mu_A)$ and $B = (\sigma_B^2, \mu_B)$, as shown in Fig. 4.4.*

Remark 125 (Efficient frontiers of two risky assets portfolios)
In the case of a generic correlation, the frontier is represented by an equation of a hyperbole in the plane (σ_P, μ_P), ranging between the three curves just described with $\rho_{AB} = \{-1, 0, 1\}$. Note that in Fig. 4.3 we indicate the part of the frontier that is inefficient with a dashed line, while we indicate the efficient frontier with a solid line.

Portfolio with n risky assets when shortsellings are allowed

We show here how to find a closed-form solution for the frontier of a feasible portfolios set for a market with n risky assets, when shortsellings are allowed. Furthermore, at the end of this section, we will describe how to numerically solve the Markowitz model when shortsellings are not allowed, using the MATLAB Optimization toolbox (see Section 3.2). We denote the expected return of the asset i by μ_i, and the covariance between returns of the assets i and j (for $i, j = 1, \ldots, n$) by σ_{ij}. The classical Mean-Variance (MV) model can be expressed

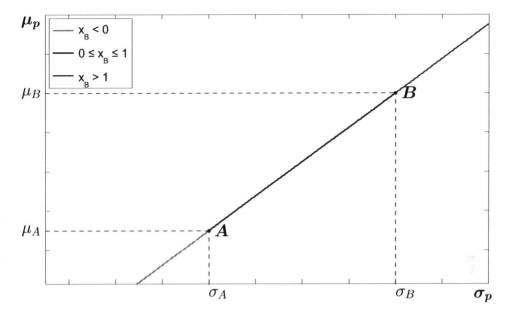

Figure 4.4: Example of a efficient frontier for two assets with perfect positive correlation when shortselling is allowed.

by the following bi-objective optimization problem with an equality constraint, representing the budget constraint

$$
\begin{cases}
\min & \sigma_P^2 = x^T \Sigma x = \displaystyle\sum_{i=1}^{n} \sum_{j=1}^{n} x_i x_j \sigma_{ij} \\
\max & \mu_P = \mu^T x = \displaystyle\sum_{i=1}^{n} \mu_i x_i \\
\text{s.t.} & \\
& u^T x = 1
\end{cases}
\tag{4.16}
$$

where, therefore, μ is the vector of expected returns, Σ is the variance-covariance matrix and u is the all-ones vector. Using, e.g., the ε-constraint method described in Section 3.3.1, such multi-objective optimization problem can be formulated as the following single-objective optimization problem

$$
\begin{cases}
\min & x^T \Sigma x \\
\text{s.t.} & \\
& \mu^T x \geq \eta \\
& u^T x = 1
\end{cases}
\quad \Rightarrow \quad
\begin{cases}
\min & x^T \Sigma x \\
\text{s.t.} & \\
& \mu^T x = \eta \\
& u^T x = 1
\end{cases}
\tag{4.17}
$$

127

where η is the required level of the portfolio expected return and the implication is due to the convexity of Problem (4.17) (see Proposition 129 and Corollary 130). Assuming that the covariance matrix Σ is positive definite, we can simply find the global minimum of Problem (4.17), searching for the stationary points of the corresponding Lagrange function. The Lagrangian associated to Problem (4.17) in matricial form is

$$L(x, \lambda_1, \lambda_2) = x^T \Sigma x - \lambda_1 \left(\mu^T x - \eta \right) - \lambda_2 \left(u^T x - 1 \right),$$

where λ_1 and λ_2 are called Lagrange multipliers. To solve this constrained minimization problem, we need to find the points for which the components of the Lagrangian gradient are equal to zero, namely the stationary points. Thus, we have

$$\begin{cases} \dfrac{\partial L}{\partial x} = 0 \\ \dfrac{\partial L}{\partial \lambda_1} = 0 \\ \dfrac{\partial L}{\partial \lambda_2} = 0 \end{cases} \Rightarrow \begin{cases} 2\Sigma x - \lambda_1 \mu - \lambda_2 u = 0 \\ \mu^T x - \eta = 0 \\ u^T x - 1 = 0 \end{cases}$$

Finding the optimal vector of weights x from the first equation, we obtain

$$\begin{cases} x = \dfrac{\lambda_1}{2} \Sigma^{-1} \mu + \dfrac{\lambda_2}{2} \Sigma^{-1} u \\ \dfrac{\lambda_1}{2} \mu^T \Sigma^{-1} \mu + \dfrac{\lambda_2}{2} \mu^T \Sigma^{-1} u = \eta \\ \dfrac{\lambda_1}{2} u^T \Sigma^{-1} \mu + \dfrac{\lambda_2}{2} u^T \Sigma^{-1} u = 1 \end{cases} \tag{4.18}$$

The last two equations can be written in matricial form as follows

$$\begin{bmatrix} \mu^T \Sigma^{-1} \mu & \mu^T \Sigma^{-1} u \\ u^T \Sigma^{-1} \mu & u^T \Sigma^{-1} u \end{bmatrix} \begin{bmatrix} \dfrac{\lambda_1}{2} \\ \dfrac{\lambda_2}{2} \end{bmatrix} = \begin{bmatrix} \eta \\ 1 \end{bmatrix}. \tag{4.19}$$

Now, denoting

$$\begin{aligned} a &= \mu^T \Sigma^{-1} \mu \\ b &= u^T \Sigma^{-1} u \\ c &= u^T \Sigma^{-1} \mu = \mu^T \Sigma^{-1} u \end{aligned}$$

one can write Expression (4.19) as follows

$$\begin{bmatrix} a & c \\ c & b \end{bmatrix} \begin{bmatrix} \dfrac{\lambda_1}{2} \\ \dfrac{\lambda_2}{2} \end{bmatrix} = \begin{bmatrix} \eta \\ 1 \end{bmatrix}. \tag{4.20}$$

Solving this system of linear equations, we obtain the optimal values $\dfrac{\lambda_1^*}{2} = \dfrac{b\eta - c}{d}$ and $\dfrac{\lambda_2^*}{2} = \dfrac{a - c\eta}{d}$, where $d = ab - c^2$. Then, we have the following closed-form solution x^* for Problem (4.17)

$$x^* = \frac{\Sigma^{-1}}{d} \left[(b\eta - c)\mu + (a - c\eta)u \right] .$$

Furthermore, since $\Sigma x^* = \dfrac{\lambda_1^*}{2}\Sigma\Sigma^{-1}\mu + \dfrac{\lambda_2^*}{2}\Sigma\Sigma^{-1}u = \dfrac{1}{2}(\lambda_1^*\mu + \lambda_2^*u)$, $x^{*T}\mu = \eta = \mu_P^*$, and $x^{*T}u = 1$, we have

$$\sigma_P^{*2} = x^{*T}\Sigma x^* = \frac{1}{2}(\lambda_1^*\eta + \lambda_2^*) = \frac{b}{d}\mu_P^{*2} - 2\frac{c}{d}\mu_P^* + \frac{a}{d}, \tag{4.21}$$

that in the Mean-Variance plane is the equation of a horizontal parabola representing the frontier of a set of feasible portfolios.

Remark 126 (Subjective choice on the efficient frontier)
Once the "objective" Mean-Variance analysis has been made, all operators have the same efficient frontier, on which to make a choice. In other words, since the efficient frontier consists of an infinite number of Pareto optimal portfolios, to choose a specific portfolio, we should consider the investor's unique preferences. As mentioned in Section 4.2.1, in the Expected Utility Theory framework a subjective preference is described by a utility function $u(w)$. Supposing that each investor has a utility (real) function, that represents his/her preferences, if w is the wealth, then $u(w)$ is the investor satisfaction associated to w. For a rational agent, the larger the wealth and the greater the satisfaction. Mathematically, this means that $u(w)$ is a monotonically increasing function, namely if $w_1 < w_2$, then $u(w_1) < u(w_2)$. If we assume that $u(w)$ is differentiable, then this condition can be obtained by requiring that the first derivative is positive, namely $u'(w) > 0$. The investor's choices are made using the criterion of maximizing the expected utility, and if the investor is risk-averse (risk-lover), then the utility function $u(w)$ is concave, i.e., $u''(w) < 0$ (convex, i.e., $u''(w) > 0$). Returning to the portfolio selection framework, suppose to represent the operators' preferences, in the (σ^2, μ) plane, by means of contour lines of the utility function $U(\sigma^2, \mu)$, i.e., curves along which the operators have a constant level of satisfaction. Thus, each investor chooses the portfolio composition that maximizes his/her satisfaction, and belongs to the efficient frontier.

Portfolio of a risk-free and a risky asset

A risk-free investment has variance $\sigma_f^2 = 0$ and a constant expected return r_f, and it can represent the possibility to borrow or lend money at the same risk-free rate r_f. Clearly, in the Mean-Variance plane we identify this asset by a point $F = \left(\sigma_f^2, \mu_f\right) = (0, r_f)$ lying on the vertical axis. Let us combine the risk-free asset F with a risky one, denoted by $A = (\sigma_A^2, \mu_A)$, where, obviously, $\sigma_A^2 > \sigma_f^2$ and $\mu_A > r_f$ (otherwise F should be preferred to A). Now, if we denote by x_A and x_f the portfolio weights of the risky and the risk-free assets, respectively, then the resulting portfolio composition will satisfy the following system of equations:

$$
\begin{cases}
1 = x_A + x_f \\
\mu_P = x_A \mu_A + x_f r_f \\
\sigma_P^2 = x_A^2 \sigma_A^2 + 2 \sigma_A \underbrace{\sigma_f}_{=0} x_A x_f + x_f^2 \underbrace{\sigma_f^2}_{=0}
\end{cases}
\implies
\begin{cases}
x_A = 1 - x_f \\
\mu_P = (1 - x_f)\mu_A + x_f r_f \\
\sigma_P = (1 - x_f)\sigma_A
\end{cases}
$$

Thus, in the Mean-Volatility plane, the (efficient) frontier is described by a straight line between the points $A = (\sigma_A, \mu_A)$ and $F = (0, r_f)$, namely

$$
\mu_P = r_f + \frac{\mu_A - r_f}{\sigma_A} \sigma_P . \tag{4.22}
$$

Furthermore, as shown in Fig. 4.5, the red line AF indicates that the investor lends money ($x_f > 0$), while on the blue line the investor borrows money ($x_f < 0$).

Portfolio of a risk-free and n risky assets

Let us consider here the case of a portfolio with a risk-free and n risky assets. The portfolio expected return $\hat{\mu}_P$ and variance $\hat{\sigma}_P^2$ lying on the frontier will satisfy the following system of equations:

$$
\begin{cases}
\hat{\mu}_P = x_f r_f + \sum_{i=1}^{n} x_i \mu_i \\
\hat{\sigma}_P^2 = \sum_{i=1}^{n} \sum_{j=1}^{n} x_i x_j \sigma_{i,j} \\
\sum_{i=1}^{n} x_i = 1 - x_f
\end{cases}
\tag{4.23}
$$

where x_i, with $i = 1, \dots, n$, is the fraction of the capital to be invested in the risky asset i, and x_f is the fraction to be invested in the risk-free asset.

130

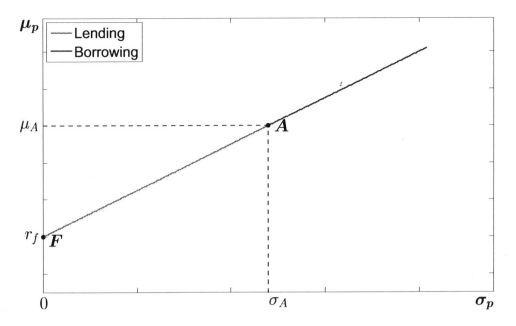

Figure 4.5: Example of an efficient frontier obtained by a combination of a risk-free and a risky asset.

Now, let us define the following new variables $y_i = \dfrac{x_i}{(1 - x_f)}$ with $i = 1, \ldots, n$. From the last equation of (4.23), this implies that $\sum\limits_{i=1}^{n} y_i = 1$. Then, substituting $x_i = y_i(1 - x_f)$ in Problem (4.23), we obtain

$$
\begin{cases}
\hat{\mu}_P = x_f r_f + (1 - x_f) \underbrace{\sum_{i=1}^{n} y_i \mu_i}_{\mu_P} \\[4mm]
\hat{\sigma}_P^2 = (1 - x_f)^2 \underbrace{\sum_{i=1}^{n} \sum_{j=1}^{n} y_i y_j \sigma_{ij}}_{\sigma_P^2} \\[4mm]
\sum_{i=1}^{n} y_i = 1
\end{cases}
\tag{4.24}
$$

where $\mu_P = \sum\limits_{i=1}^{n} y_i \mu_i$ and $\sigma_P^2 = \sum\limits_{i=1}^{n} \sum\limits_{j=1}^{n} y_i y_j \sigma_{ij}$ are respectively the expected

131

return and variance of a Markowitz portfolio $(y_1, y_2, \ldots, y_n)$ composed only by risky assets, whose weights sum to one.

From the second equation of (4.24) we can write $x_f = 1 - \dfrac{\hat{\sigma}_P}{\sigma_P}$. Thus, it is straightforward to see that in the Mean-Volatility plane the relation between $\hat{\mu}_P$ and $\hat{\sigma}_P$ is a straight line, similarly to the case of a portfolio with a risk-free and a risky assets (see Eq. (4.22)):

$$\hat{\mu}_P = r_f + \frac{\mu_P - r_f}{\sigma_P}\hat{\sigma}_P \ . \tag{4.25}$$

As a consequence, the efficient frontier of a portfolio with a risk-free and n risky assets can be obtained by maximizing the slope $\theta = \dfrac{\mu_P - r_f}{\sigma_P}$ of Equation (4.25) as follows

$$\begin{cases} \max \quad \theta = \dfrac{\mu_P - r_f}{\sigma_P} = \dfrac{\sum\limits_{i=1}^{n} y_i \mu_i - r_f}{\sqrt{\sum\limits_{i=1}^{n}\sum\limits_{j=1}^{n} \sigma_{ij} y_i y_j}} \\[2em] \text{s.t.} \\ \qquad \sum\limits_{i=1}^{n} y_i = 1 \end{cases} \tag{4.26}$$

As highlighted in Fig. 4.6, the optimal value of Problem (4.26) is $\theta^* = \dfrac{\mu_M - r_f}{\sigma_M}$, where the point $M = (\sigma_M, \mu_M)$, named *Market* portfolio, is the point of tangency between the straight line (4.25) and the Markowitz efficient frontier composed only of n risky assets. In other words, the efficient frontier of a portfolio with a risk-free and n risky assets is represented by the following straight line passing through the points $F = (0, r_f)$ and $M = (\sigma_M, \mu_M)$

$$\hat{\mu}_P^* = r_f + \frac{\mu_M - r_f}{\sigma_M}\hat{\sigma}_P^*,$$

called Capital Market Line (CML). Note that $\theta^* = \dfrac{\mu_M - r_f}{\sigma_M}$ can also be interpreted as the market price of risk. Indeed, it provides the increment of the return corresponding to a unit increment of risk $\Delta\hat{\sigma}_P^*$ (in terms of volatility) for the efficient portfolios. In other words, θ^* is the risk premium per unit of risk assumed by an investor, along CML.

Remark 127 (Finding the Market portfolio) *Since Problem (4.26) is a nonconvex optimization problem, it can be difficult to solve it. However, it can be*

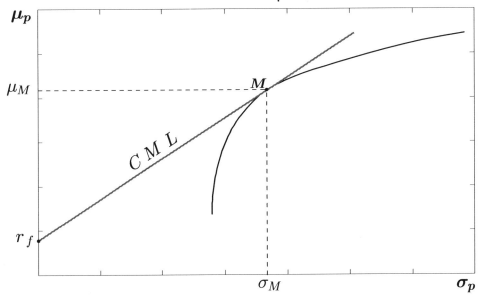

Figure 4.6: Capital Market Line

transformed in an equivalent problem that is convex (see, e.g., Cornuejols and Tütüncü, 2006). Indeed, assuming that there exists a portfolio $\hat{y} \in C$ such that $\mu^T \hat{y} > r_f$, where C is the set of feasible portfolios with $\sum_{i=1}^{n} y_i = 1$ (e.g., C can also include the no shortselling constraints, $y_i \geq 0$ for all i), we could define the following new variables:

$$\mu^T y - r_f = \frac{1}{\xi} \quad and \quad y\xi = z.$$

Then, the inverse of the objective function of Problem (4.26) becomes

$$\frac{\sigma_P}{\mu_P - r_f} = \frac{\sqrt{y^T \Sigma y}}{\mu^T y - r_f} = \sqrt{(y\xi)^T \Sigma y\xi} = \sqrt{z^T \Sigma z}.$$

Furthermore, since $u^T y = 1$ (where u is a vector of ones),

$$\mu^T y - r_f = \mu^T y - u^T y r_f = (\mu - r_f u)^T y = \frac{1}{\xi}$$
$$\Rightarrow (\mu - r_f u)^T y\xi = 1 \Rightarrow (\mu - r_f u)^T z = 1$$

and, since $y\xi = z$, $u^T y = u^T \frac{z}{\xi} = 1$. Thus, the Market portfolio y^* can be found solving the following (equivalent) convex problem

$$
\begin{cases}
\min \quad z^T \Sigma z \\
s.t. \\
\quad (\mu - r_f u)^T z = 1 \\
\quad u^T \frac{z}{\xi} = 1 \\
\quad \xi \geq 0, \ z \geq 0 \\
\quad (z, \xi) \in C^+
\end{cases}
\tag{4.27}
$$

where $\mathbb{R}^{n+1} \supseteq C^+ = \{z \in \mathbb{R}^n, \xi > 0 : \frac{z}{\xi} \in C\} \cup (0,0)$. If the pair (z^*, ξ^*) is the solution of (4.27), then $y^* = \frac{z^*}{\xi^*}$ is the solution of Problem (4.26) (see also Stoyanov et al, 2007).

Problem (4.27) is a convex Quadratic Programming (QP) problem which can be solved by the MATLAB built-in function `quadprog` (see Section 3.2.2), as shown in Exercise 133.

A second heuristic method to find the Market portfolio is to roughly apply the Markowitz model on a set of assets that also contain the risk free one. Thus, the resulting problem is

$$
\begin{cases}
\min \quad \tilde{x}^T \tilde{\Sigma} \tilde{x} \\
s.t. \\
\quad \tilde{\mu}^T \tilde{x} = \eta \\
\quad u^T \tilde{x} = 1 \\
\quad \tilde{x} \geq 0
\end{cases}
\tag{4.28}
$$

where $\tilde{x}^T = \{x_1, \ldots, x_N, x_f\}$, $\tilde{\Sigma} = \begin{bmatrix} \Sigma & 0 \\ 0 & 0 \end{bmatrix}$ and $\tilde{\mu}^T = \{\mu_1, \ldots, \mu_n, r_f\}$. The Market portfolio $M = (\sigma_M, \mu_M)$ corresponds to the point of the frontier where $x_f = 0$ and $x_i > 0$ for at least an asset i.

Portfolio with n risky assets when no shortsellings are allowed

We propose here a numerical technique to solve the Markowitz model when short sales are not allowed. In this case, the MV model becomes

$$
\left\{
\begin{array}{ll}
\min & \sum\limits_{i=1}^{n}\sum\limits_{j=1}^{n}\sigma_{ij}x_ix_j \\
\text{s.t.} & \\
& \sum\limits_{i=1}^{n}\mu_ix_i = \eta \\
& \sum\limits_{i=1}^{n}x_i = 1 \\
& x_i \geq 0 \qquad i = 1,\ldots,n
\end{array}
\right.
\qquad \Leftrightarrow \qquad
\left\{
\begin{array}{ll}
\min & x^T\Sigma x \\
\text{s.t.} & \\
& \mu^T x = \eta \\
& u^T x = 1 \\
& x \geq 0
\end{array}
\right.
\qquad (4.29)
$$

where, as in Problem (4.17), η is the required level of the portfolio expected return. Problem (4.29) is a convex Quadratic Programming (QP) problem which can only be solved numerically by a number of efficient algorithms with moderate computational effort, even for large n.

As shown by the generic QP problem (3.6), to solve Problem (4.29) by means of the built-in Function `quadprog`, we can set its parameters as follows

$$
H = 2\Sigma = 2\begin{bmatrix}
\sigma_{11} & \sigma_{12} & \cdots & \sigma_{1n} \\
\sigma_{21} & \sigma_{22} & \cdots & \sigma_{2n} \\
\vdots & \vdots & \ddots & \vdots \\
\sigma_{n1} & \sigma_{n2} & \cdots & \sigma_{nn}
\end{bmatrix};
$$

$$
f = \begin{pmatrix} 0 \\ 0 \\ \vdots \\ 0 \end{pmatrix}; \quad
x = \begin{pmatrix} x_1 \\ x_2 \\ \vdots \\ x_n \end{pmatrix}; \quad
l_b = \begin{pmatrix} 0 \\ 0 \\ \vdots \\ 0 \end{pmatrix}; \quad
u_b = \begin{pmatrix} +\infty \\ +\infty \\ \vdots \\ +\infty \end{pmatrix};
$$

$$
b_{eq} = \begin{pmatrix} \eta \\ 1 \end{pmatrix}; \quad
A_{eq} = \begin{pmatrix} \mu_1 & \mu_2 & \cdots & \mu_n \\ 1 & 1 & \cdots & 1 \end{pmatrix}.
$$

We denote by $\tilde{\sigma}_P^2(\eta)$ the optimal value of (4.29) as a function of η. Let η_{min} denote the value of the portfolio expected return $\sum_{i=1}^{n}\mu_ix_i$ at an optimal solution of the problem obtained by deleting the first constraint in (4.29), and let $\eta_{max} = \max\{\mu_1,\ldots,\mu_n\}$ (see Remark 128). Then, the graph of $\tilde{\sigma}_P^2(\eta)$ on the interval $[\eta_{min},\eta_{max}]$ contains the risk-return values of all the efficient portfolios (see Cesarone et al, 2013), and is usually approximated by solving Problem (4.29) for several (equally spaced) values of $\eta \in [\eta_{min},\eta_{max}]$.

Remark 128 (The portfolio with maximum expected return) *Since in Problem (4.29) we have the budget and the no short sales constraints, the portfolio weights $x_i \in [0,1]$ for all i. Then, the maximum value assumed by the*

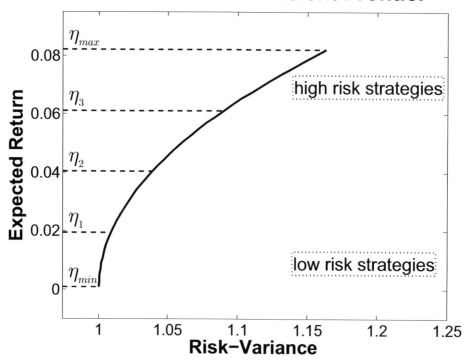

Figure 4.7: Example of a Risk-Return *efficient frontier* for different levels of target returns

target level of the portfolio expected return η will be

$$\eta = \sum_{i=1}^{n} \mu_i x_i \leq \sum_{i=1}^{n} \eta_{max} x_i = \eta_{max} \sum_{i=1}^{n} x_i = \eta_{max} \, ,$$

where $\eta_{max} = \max\{\mu_1, \ldots, \mu_n\}$ and, therefore, it corresponds to the target level of the portfolio with only one asset, namely the asset with the greatest expected return.

Proposition 129 (Cesarone et al (2013)) *The convexity of Problem (4.29) implies that, for $\eta \in [\eta_{min}, \eta_{max}]$, the function $\tilde{\sigma}_P^2(\eta)$ is increasing and convex.*

Proof. Let $\eta_0, \eta_1 \in [\eta_{min}, \eta_{max}]$, and let x_0, x_1 denote the two corresponding solutions of Problem (4.29). Then, due to the linearity of expected return (see

Remark 85), for any $\lambda \in [0,1]$ $x_\lambda = (1-\lambda)x_0 + \lambda x_1$ is a feasible solution to (4.29) for $\eta_\lambda = (1-\lambda)\eta_0 + \lambda \eta_1$. The convexity of $\tilde{\sigma}_P^2(\eta)$ follows from the convexity of the variance $\sigma_P^2(x) = x^T \Sigma x$ (see Remark 121):

$$\tilde{\sigma}_P^2(\eta_\lambda) \leq \sigma_P^2(x_\lambda) = \sigma_P^2((1-\lambda)x_0 + \lambda x_1) \leq (1-\lambda)\sigma_P^2(x_0) + \lambda \sigma_P^2(x_1)$$

Thus

$$\tilde{\sigma}_P^2(\eta_\lambda) \leq (1-\lambda)\tilde{\sigma}_P^2(\eta_0) + \lambda \tilde{\sigma}_P^2(\eta_1)$$

To prove isotonicity of $\tilde{\sigma}_P^2$, take any $\eta_0 < \eta_1 \in [\eta_{min}, \eta_{max}]$. Then, for some $\lambda \in (0,1)$, we have $\eta_0 = \lambda \eta_{min} + (1-\lambda)\eta_1$, so that

$$\tilde{\sigma}_P^2(\eta_0) \leq \lambda \tilde{\sigma}_P^2(\eta_{min}) + (1-\lambda)\tilde{\sigma}_P^2(\eta_1) \leq \lambda \tilde{\sigma}_P^2(\eta_1) + (1-\lambda)\tilde{\sigma}_P^2(\eta_1) = \tilde{\sigma}_P^2(\eta_1),$$

where the last inequality follows from the fact that $\tilde{\sigma}_P^2(\eta_{min}) \leq \tilde{\sigma}_P^2(\eta)$ for all $\eta \in [\eta_{min}, \eta_{max}]$, by definition of η_{min}. ∎

From the above proposition, we immediately derive the following result.

Corollary 130 (Cesarone et al (2013)) *For $\eta \in [\eta_{min}, \eta_{max}]$ the solution of Problem (4.29) does not change if we replace the expected return constraint with $\sum_{i=1}^n \mu_i x_i \geq \eta$.*

In the following exercises, we show *(i)* how to determine the efficient frontier related to the Mean-Variance model on a real market; *(ii)* how to obtain the frontiers of a portfolio with two risky assets for different levels of correlation (see Fig. 4.3); *(iii)* how to compute the Capital Market Line (see Fig. 4.6).

Exercise 131 (Markowitz portfolios) *Write a Script* S_Markowitz.m *that solves the following points.*

1. *Import the data from* weekly_EUROSTOXX50_price_time.txt *defining a matrix* **A**. *Note that this file contains time series of the weekly prices for the EUROSTOXX50 market. Its first column is a vector of dates in the MATLAB numerical format, while the rest of columns are the prices of the assets. Each row is the price for a specified date.*

2. *Define a column vector* times *that includes the first column of* **A**, *and create a matrix* **P** *that contains the assets prices.*

3. *For each asset, compute the time series of the linear returns and save them in a matrix called* **RR**.

4. *Compute the assets expected returns vector μ and the covariance matrix Σ.*

137

5. *Compute the return of the minimum risk portfolio (η_{min}) and that of the maximum return (η_{max}).*

6. *Compute the vector η of target portfolio expected returns considering that η has to contain 100 equally-spaced values between η_{min} and η_{max} (eta=linspace(eta_min,eta_max,100)).*

7. *For each value of η, compute the optimal Markowitz portfolio, saving the optimal value of risk in the sigma_2 vector.*

   ***Hint**: to obtain a smoother efficient frontier use the optimset function thus improving the accuracy of the optimal solutions. More precisely, set MaxIter=1.e7, TolFun=1.e-10 and TolX=1.e-10.*

8. *Save RR, the vector η of target returns, and the sigma_2 vector in a workspace called MarkRiskyAssets.mat.*

9. *Finally, make the graph of the target expected returns eta as a function of sigma_2, and then save it as Markowitz_EF.jpg.*

Sol.: See Script S_Markowitz.

Exercise 132 (Two risky assets portfolio) *Solve the following points in the Script S_TwoRiskyAssets.m.*

1. *Import the data from weekly_EUROSTOXX50_price_time.txt, creating a matrix A as in Exercise 131. Then, define a column vector times that includes the first column of A, and create a matrix of prices P containing only the 6th and 7th column of A.*

2. *Compute the historical linear returns (Ret) from P.*

3. *Supposing to have a parametric correlation cor, compute the frontiers determined by the two assets using 100 equally-spaced values of the portfolio target return $\eta \in [\min\{\mu_1, \ldots, \mu_n\}, \max\{\mu_1, \ldots, \mu_n\}]$, by varying cor between -1 and 1.*

4. *As in Fig. 4.3, in the Mean-Volatility plane, plot the frontiers obtained by varying η and cor, and save them on the file TwoRiskyAssets.tif.*

Sol.: See Script S_TwoRiskyAssets.

Exercise 133 (Market portfolio and Capital Market Line)
Write a Script S_CML.m *that solves the following points.*

1. *Load* MarkRiskyAssets.mat *that contains the variables of Exercise 131.*

2. *Compute the assets expected returns vector μ and the covariance matrix Σ, define a risk-free investment with return* r_f=0.0025, *and find the Market Portfolio $M = (\sigma_M, \mu_M)$ by solving the Quadratic Programming problem (4.27).*

3. *Define an anonymous function (see Section 1.1.6) representing the Capital Market Line equation*

$$\sigma_P = \frac{\sigma_M}{\mu_M - r_f}\mu_P - \frac{r_f \sigma_M}{\mu_M - r_f} .$$

4. *Finally, in the same graph, plot the Markowitz efficient frontier (black), the CML (red) up to a level of expected return equal to η_{max}, and mark the Market portfolio M by a point. Then, save it as* CML.jpg *(see Fig. 4.6).*

Sol.: See Script S_CML.

4.2.4 Effects of diversification for an EW portfolio

The concept of diversification can be qualitatively related to the portfolio risk reduction due to the process of compensation caused by the co-movement among assets that leads to a potential attenuation of the exposure to risk determined by individual asset shocks. However, the portfolio diversification can be a sophisticated concept to formalize, and the question of which measure of diversification is most appropriate is still open (see, e.g., Meucci, 2009a; Lhabitant, 2017). Nevertheless, it is possible to provide an idea of how the portfolio risk depends on diversification, by considering an Equally Weighted (EW) portfolio, i.e., a portfolio in which each asset is purchased with the same relative weight, $x_i = 1/n$ with $i = 1, \ldots n$, where n is the number of assets in the market considered. In the same framework of the Markowitz model, let us consider the variance of such a portfolio

$$\sigma_{EW}^2 = \frac{1}{n^2} \sum_{i=1}^{n} \sum_{j=1}^{n} \sigma_{ij}$$

$$= \underbrace{\frac{1}{n^2} \sum_{i=1}^{n} \sigma_i^2}_{\text{Variance term}} + \underbrace{\frac{1}{n^2} \sum_{i=1}^{n} \sum_{\substack{j=1 \\ j \neq i}}^{n} \sigma_{ij}}_{\text{Covariance term}} . \qquad (4.30)$$

Let us denote the average of the assets variances by $\overline{\sigma_V^2} = \frac{1}{n} \sum_{i=1}^{n} \sigma_i^2$ and the

average of their covariances by $\overline{\sigma_C^2} = \frac{1}{n^2 - n} \sum_{i=1}^{n} \sum_{\substack{j=1 \\ j \neq i}}^{n} \sigma_{ij}$. Then we can rewrite

(4.30) as

$$\sigma_{EW}^2 = \frac{1}{n} \overline{\sigma_V^2} + \frac{n^2 - n}{n^2} \overline{\sigma_C^2}$$

$$= \frac{1}{n} \overline{\sigma_V^2} + \left(1 - \frac{1}{n}\right) \overline{\sigma_C^2} .$$

Therefore, with some abuse of the notation, we can say that

$$\sigma_{EW}^2 \xrightarrow[n \to +\infty]{} \overline{\sigma_C^2} .$$

This means that, when increasing the number of assets selected in the portfolio, the *Variance term* approaches zero, and the main contribution to the total portfolio variance is due to the *Covariance term*. This phenomenon can be empirically shown by the following famous experiment.

Exercise 134 (Fama's Experiment) *Given the returns of n assets in an investment universe, we can define the $n \times n$ covariance matrix Σ. To examine the portfolio diversification effects, one can study the volatility of the portfolio constructed investing all capital in the first asset, namely $\sigma(1) = \sqrt{1\sigma_{11}1}$. Then, it is possible to repeat the procedure calculating the volatility of the Equally Weighted portfolio consisting of the first two assets, where, therefore, the vector of weights is $x = (1/2, 1/2)$. This procedure can be iterated until the number of assets selected is equal to n. In a Script named* S_Fama, *import the data contained in* weekly_MIBTEL_price_time.txt, *defining a matrix* A. *Furthermore, define a column vector* times *that includes the*

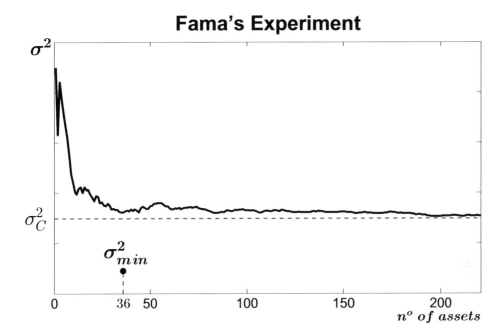

Fama's Experiment

Figure 4.8: Effects of diversification on the EW portfolio variance by increasing the number of the assets included

first column of A, *and create a matrix* P *containing the remaining columns, i.e., the assets prices. Compute the time series of linear returns in a matrix* RR, *and their covariance matrix* Sigma. *Save this last matrix in a .txt file. Then, calculate the* $1 \times n$ *vector* sigma_EW *whose k-th element is the volatility of the Equally Weighted portfolio obtained by the first k rows and columns of* Σ, *with* $k = 1, \dots, n$. *Finally, graph* sigma_EW, *where the cardinality of the Equally Weighted portfolio is on the horizontal axis, and save it as* Fama_exp.tif.

Sol.: See Script S_Fama.

4.2.5 Mean-Mean Absolute Deviation model

A further symmetric risk measure often used in the literature is the Mean Absolute Deviation (MAD). As in Konno and Yamazaki (1991), MAD is defined as the expected value of the absolute deviation of the portfolio return $R_P(x)$ from its mean $\mu_P(x)$:

$$MAD(x) = E[|R_P(x) - \mu_P(x)|] = E\left[\left|\sum_{i=1}^{n} R_i x_i - \sum_{i=1}^{n} \mu_i x_i\right|\right] \qquad (4.31)$$

141

where R_i is the random return of asset i and μ_i is its expected value. The risk-return analysis based on MAD as risk measure can be obtained following the same steps as for the Markowitz model (see Problems (4.16) and (4.17)). Then, the Mean-MAD model can be expressed by the following bi-objective optimization problem:

$$\begin{cases} \min \quad MAD(x) = E\left[\left|\sum_{i=1}^{n} R_i x_i - \sum_{i=1}^{n} \mu_i x_i\right|\right] \\ \max \quad \mu_P(x) = \sum_{i=1}^{n} \mu_i x_i \\ \text{s.t.} \\ \qquad u^T x = 1 \\ \qquad x \geq 0 \end{cases} \tag{4.32}$$

where u is the all-ones vector. Using the ε-constraint method (see Section 3.3.1), the multi-objective optimization problem (4.32) can be formulated as the following single-objective optimization problem

$$\begin{cases} \min \quad MAD(x) \\ \text{s.t.} \\ \qquad \mu^T x \geq \eta \\ \qquad u^T x = 1 \\ \qquad x \geq 0 \end{cases} \quad \Rightarrow \quad \begin{cases} \min \quad MAD(x) \\ \text{s.t.} \\ \qquad \mu^T x = \eta \\ \qquad u^T x = 1 \\ \qquad x \geq 0 \end{cases} \tag{4.33}$$

where μ is the vector of the assets expected returns, η is the required level of the portfolio expected return, and the implication is due to the convexity of Problem (4.33).

It is worth mentioning here that we denote by R_i the random variable representing the i-th assets return, while by $r_{i,t}$ the historical realization of that r.v. at time t. Having said that and assuming (as usual in Portfolio Selection) that all historical scenarios are equally likely, in the discrete case Problem (4.33) can be formulated as follows:

$$\begin{cases} \min_{x} \quad \frac{1}{T} \sum_{t=1}^{T} \left|\sum_{i=1}^{n} r_{i,t} x_i - \sum_{i=1}^{n} \mu_i x_i\right| \\ \text{s.t.} \\ \qquad \sum_{i=1}^{n} \mu_i x_i = \eta \\ \qquad \sum_{i=1}^{n} x_i = 1 \\ \qquad x_i \geq 0 \qquad\qquad i = 1, \ldots, n \end{cases} \tag{4.34}$$

As in Konno and Yamazaki (1991), we linearize Problem (4.34) by introducing T auxiliary variables d_t defined as the absolute deviation of the portfolio return from its mean. Therefore, we substitute $|\sum_{i=1}^{n} r_{i,t}x_i - \sum_{i=1}^{n} \mu_i x_i| = d_t$ by adding the following constraints: $d_t \geq 0$, $d_t \geq \sum_{i=1}^{n}(r_{i,t} - \mu_i)x_i$ and $d_t \geq -\sum_{i=1}^{n}(r_{i,t} - \mu_i)x_i$. Hence, the Mean-MAD model can be rewritten as the following Linear Programming (LP) problem:

$$
\begin{cases}
\min_{(x,d)} & \frac{1}{T}\sum_{t=1}^{T} d_t \\
\text{s.t.} & \\
& d_t \geq \sum_{i=1}^{n}(r_{i,t} - \mu_i)x_i \quad t = 1, \ldots, T \\
& d_t \geq -\sum_{i=1}^{n}(r_{i,t} - \mu_i)x_i \quad t = 1, \ldots, T \\
& d_t \geq 0 \quad t = 1, \ldots, T \\
& \sum_{i=1}^{n} \mu_i x_i = \eta \\
& \sum_{i=1}^{n} x_i = 1 \\
& x_i \geq 0 \quad i = 1, \ldots, n
\end{cases}
\tag{4.35}
$$

This problem is an LP with $n+T$ variables and $3T+n+2$ constraints, where the portfolio weights $(x_1, \ldots, x_n)$ and the auxiliary variables $(d_1, \ldots, d_T)$ represent the decision variables of Problem (4.35).

The MAD model assumes no particular distribution for asset returns. In fact, it has been applied where there are asymmetric return distributions. However, as shown by Konno and Yamazaki (1991), MAD is a risk measure equivalent to variance under the assumption of multivariate normal returns.

Now, we describe how to use the built-in MATLAB function `linprog` for implementing the Mean-MAD model. As shown by the generic LP problem (3.4), to solve Problem (4.35) we can set its parameters as follows

$$
A = \begin{bmatrix}
r_{1,1} - \mu_1 & r_{1,2} - \mu_2 & \cdots & r_{1,n} - \mu_n & -1 & 0 & \cdots & 0 \\
r_{2,1} - \mu_1 & r_{2,2} - \mu_2 & \cdots & r_{2,n} - \mu_n & 0 & -1 & \cdots & 0 \\
\vdots & \vdots & \ddots & \vdots & \vdots & \vdots & \ddots & \vdots \\
r_{T,1} - \mu_1 & r_{T,2} - \mu_2 & \cdots & r_{T,n} - \mu_n & 0 & 0 & \cdots & -1 \\
-(r_{1,1} - \mu_1) & -(r_{1,2} - \mu_2) & \cdots & -(r_{1,n} - \mu_n) & -1 & 0 & \cdots & 0 \\
-(r_{2,1} - \mu_1) & -(r_{2,2} - \mu_2) & \cdots & -(r_{2,n} - \mu_n) & 0 & -1 & \cdots & 0 \\
\vdots & \vdots & \ddots & \vdots & \vdots & \vdots & \ddots & \vdots \\
-(r_{T,1} - \mu_1) & -(r_{T,2} - \mu_2) & \cdots & -(r_{T,n} - \mu_n) & 0 & 0 & \cdots & -1
\end{bmatrix};
$$

143

Mean-MAD Efficient Frontier

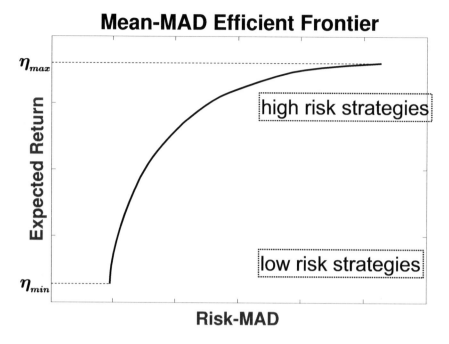

Figure 4.9: Example of the Mean-MAD *efficient frontier* for different levels of target portfolio expected returns

$$f = \frac{1}{T} \begin{pmatrix} 0 \\ \vdots \\ 0 \\ 1 \\ \vdots \\ 1 \end{pmatrix} \quad ; \qquad x = \begin{pmatrix} x_1 \\ \vdots \\ x_n \\ d_1 \\ \vdots \\ d_T \end{pmatrix} \quad ; \qquad b = \begin{pmatrix} 0 \\ \vdots \\ 0 \\ 0 \\ \vdots \\ 0 \end{pmatrix} \quad ;$$

$$l_b = \begin{pmatrix} 0 \\ \vdots \\ 0 \\ 0 \\ \vdots \\ 0 \end{pmatrix} \quad ; \qquad u_b = \begin{pmatrix} +\infty \\ \vdots \\ +\infty \\ +\infty \\ \vdots \\ +\infty \end{pmatrix} \quad ;$$

$$b_{eq} = \begin{pmatrix} \eta \\ 1 \end{pmatrix} \quad ; \qquad A_{eq} = \begin{pmatrix} \mu_1 & \mu_2 & \dots & \mu_n & 0 & \dots & 0 \\ 1 & 1 & \dots & 1 & 0 & \dots & 0 \end{pmatrix} .$$

In the following exercise, we show how to determine the efficient frontier related to the Mean-MAD model on a real market.

Exercise 135 (Mean-MAD portfolios) *Write the Script* S_MAD *that solves the following points.*

1. *Import data of* `weekly_EUROSTOXX50_price_time.txt` *in a matrix* A. *Note that, in this file, there are the weekly historical prices of the EuroStoxx 50 market index components. The first column of* A *is a vector of dates in the MATLAB numerical format, while the others are the prices of the assets. Therefore, define the* `dates` *vector considering only the first column of* A, *and a matrix* P *considering the remaining columns.*

2. *Calculate the linear returns from* P *creating a matrix* RR.

3. *Compute the assets expected returns vector* μ.

4. *Calculate the return of the Minimum-Risk portfolio* (η_{min}) *and that of the Maximum-Return portfolio* (η_{max}).

5. *Compute the* $1 \times N$ *vector* `eta` *of the target portfolio expected returns, where* $N = 100$. *Note that* $\eta \in [\eta_{min}, \eta_{max}]$.

6. *For each value of* η, *calculate the Mean-MAD optimal portfolio, and save the optimal risk values in the vector* Risk_MAD.

7. *Make the graph of the portfolio expected return as a function of risk of the optimal portfolios obtained from the previous point. Save the graph as* MAD_frontier.tif *(see Fig. 4.9).*

Sol.: See Script S_MAD.

4.2.6   Mean-Maximum Loss model

The Mean-Variance and Mean-MAD models are portfolio selection models based on symmetric risk measures (Mitra et al, 2003). Here we present the Mean-Maximum Loss model (also called the *MinMax* model), introduced by Young (1998), which uses an asymmetric measure of risk. This asymmetric measure is the portfolio maximum loss achieved in the past, where usually losses are defined as negative returns. The goal of this models is then to minimize the maximum loss (MaxLoss) w.r.t. the usual constraints on the portfolio expected return, budget, and short selling (if provided). Clearly, an alternative formulation of this problem is the maximization of the portfolio minimum return. Note that the Mean-MaxLoss model uses the norm L_∞ to measure risk, which represents a strong aversion to downside risk.

Given a time horizon $[1, \ldots, T]$, let l_P^{max} be the maximum loss of a portfolio, defined as follows

$$l_P^{max}(x) = \max_{1 \le t \le T} -\sum_{i=1}^{n} x_i r_{i,t} = -\min_{1 \le t \le T} \sum_{i=1}^{n} x_i r_{i,t} = -r_P^{min}(x), \quad (4.36)$$

where n are the available assets in the market, $R_{P,t}(x) = \sum_{i=1}^{n} x_i r_{i,t}$ is the portfolio return at time t, and $r_P^{min}(x)$ represents the portfolio minimum return. Then, the Mean-MaxLoss model can be expressed by the following bi-objective optimization problem:

$$\begin{cases} \min & l_P^{max}(x) \\ \max & \mu_P(x) = \mu^T x \\ \text{s.t.} \\ & u^T x = 1 \\ & x \ge 0 \end{cases} \quad (4.37)$$

where μ is the vector of expected returns of the assets and u is the all-ones vector. Using the ε-constraint method (see Section 3.3.1), the multi-objective optimization problem (4.37) can be formulated as the following single-objective optimization problems

$$\begin{cases} \min_{x} & \max_{1 \le t \le T} -R_{P,t}(x) \\ \text{s.t.} \\ & \mu^T x = \eta \\ & u^T x = 1 \\ & x \ge 0 \end{cases} \quad \Leftrightarrow \quad \begin{cases} -\max_{x} & \min_{1 \le t \le T} R_{P,t}(x) \\ \text{s.t.} \\ & \mu^T x = \eta \\ & u^T x = 1 \\ & x \ge 0 \end{cases} \quad (4.38)$$

where η is the required level of expected portfolio return and the equivalence of these two problems is due to the following considerations:

$$\begin{aligned} \min_{x} \max_{1 \le t \le T} -R_{P,t}(x) &= \min_{x} -\min_{1 \le t \le T} R_{P,t}(x) \\ &= -\max_{x} \min_{1 \le t \le T} R_{P,t}(x) . \end{aligned}$$

Then, the Mean-MaxLoss model can be expressed by the following problem

$$\begin{cases} -\max_{x} & \min_{1 \le t \le T} \sum_{i=1}^{n} x_i r_{i,t} \\ \text{s.t.} \\ & \sum_{i=1}^{n} \mu_i x_i = \eta \\ & \sum_{i=1}^{n} x_i = 1 \\ & x_i \ge 0 \qquad i = 1, \ldots, n \end{cases} \quad (4.39)$$

146

Note that the objective function is non-linear. However, one can linearize it just by introducing an auxiliary variable $d = \min\limits_{1 \le t \le T} \sum\limits_{i=1}^{n} x_i r_{i,t}$ with the constraint $d \le \sum\limits_{i=1}^{n} x_i r_{i,t}$. Model (4.39) then becomes

$$
\begin{cases}
\underset{(x,d)}{-\max} \quad d \\
\text{s.t.} \\
\qquad d - \sum\limits_{i=1}^{n} x_i r_{i,t} \le 0 \quad t = 1, \ldots, T \\
\qquad \sum\limits_{i=1}^{n} \mu_i x_i = \eta \\
\qquad \sum\limits_{i=1}^{n} x_i = 1 \\
\qquad x_i \ge 0 \qquad\qquad i = 1, \ldots, n
\end{cases}
\Leftrightarrow
\begin{cases}
\underset{(x,d)}{\min} \quad -d \\
\text{s.t.} \\
\qquad d - \sum\limits_{i=1}^{n} x_i r_{i,t} \le 0 \quad \forall t \\
\qquad \sum\limits_{i=1}^{n} \mu_i x_i = \eta \\
\qquad \sum\limits_{i=1}^{n} x_i = 1 \\
\qquad x_i \ge 0 \qquad\qquad \forall i
\end{cases}
$$

$$(4.40)$$

Since, by default, the built-in Function `linprog` minimizes the objective function as indicated in (3.4), we can set its parameters as follows:

$$
A = \begin{bmatrix}
-r_{1,1} & -r_{2,1} & \cdots & -r_{n,1} & 1 \\
-r_{1,2} & -r_{2,2} & \cdots & -r_{n,2} & 1 \\
\vdots & \vdots & \ddots & \vdots & \vdots \\
-r_{1,T} & -r_{2,T} & \cdots & -r_{n,T} & 1
\end{bmatrix} \quad ;
$$

$$
b = \begin{pmatrix} 0 \\ 0 \\ \vdots \\ 0 \end{pmatrix} \quad ; \quad
x = \begin{pmatrix} x_1 \\ \vdots \\ x_n \\ d \end{pmatrix} \quad ; \quad
f = \begin{pmatrix} 0 \\ \vdots \\ 0 \\ -1 \end{pmatrix} \quad ;
$$

$$
l_b = \begin{pmatrix} 0 \\ \vdots \\ 0 \\ -\infty \end{pmatrix} \quad ; \quad
u_b = \begin{pmatrix} +\infty \\ \vdots \\ +\infty \\ +\infty \end{pmatrix} \quad ;
$$

$$
b_{eq} = \begin{pmatrix} \eta \\ 1 \end{pmatrix} \quad ; \quad
A_{eq} = \begin{pmatrix} \mu_1 & \mu_2 & \cdots & \mu_n & 0 \\ 1 & 1 & \cdots & 1 & 0 \end{pmatrix} .
$$

In the following exercise, we show how to determine the efficient frontier related to the Mean-MaxLoss model on a real market.

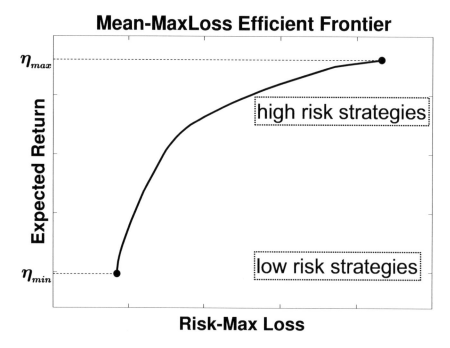

Mean-MaxLoss Efficient Frontier

η_{max} ⋯⋯⋯⋯⋯⋯⋯⋯⋯⋯⋯⋯⋯

Expected Return

high risk strategies

low risk strategies

η_{min} ⋯⋯⋯

Risk-Max Loss

Figure 4.10: Example of the Mean-MaxLoss *efficient frontier* for different levels of target portfolio expected returns

---

Exercise 136 (Mean-MaxLoss portfolios) *Write a Script that solves the following points*

1. *Import data from* `weekly_EUROSTOXX50_price_time.txt` *in a matrix* **A**. *Note that, in this file, there are the weekly historical prices of the assets that compose the market index. The first column is a vector of dates in the MATLAB numerical format, while the others are the prices. Therefore, define the* `dates` *vector considering only the first column of* **A** *and a matrix* **P** *considering the remaining columns.*

2. *Compute the linear returns from* **P** *creating a matrix* **RR** .

3. *Compute the vector* μ *of the assets expected returns.*

4. *Compute the return of the Minimum-Risk portfolio* (η_{min}) *and that of the Maximum-Return portfolio* (η_{max}) .

5. *Compute the* $1 \times N$ *vector* `eta` *of the target portfolio expected returns, where* $N = 100$. *Note that* $\eta \in [\eta_{min}, \eta_{max}]$.

148

> 6. *For each value of* `eta`, *calculate the Mean-MaxLoss optimal portfolio, and save the optimal risk values in the vector* `Risk_MaxLoss`.
>
> 7. *Plot the efficient frontier in the Return-Risk plane (namely* `eta` *vs.* `Risk_MaxLoss`*). Save the graph as* `MaxLossEF.jpg` *(see Fig. 4.10).*
>
> **Sol.:** See Script S_MaxLoss.

4.2.7 Value-at-Risk

The Value-at-Risk (VaR) is one of the most important risk management tools in the financial industry and it is commonly used in banking (see Morgan, 1996). More precisely, VaR_ε is defined as the maximum loss at a given confidence level related to a predefined time horizon T. The confidence level, which is usually 90%, 95% or 99%, is generally equal to $(1-\varepsilon)100\%$. Hence, denoting by $x = (x_1, x_2, \ldots, x_n)^T$ the vector of the assets weights in a portfolio, $VaR_\varepsilon(x)$ is the value such that the possible portfolio loss $L_P(x) = -R_P(x) = -\sum_{i=1}^n R_i x_i$ exceeds $VaR_\varepsilon(x)$ with a probability of $\varepsilon 100\%$, where, for instance, $\varepsilon = 10\%$, 5% and 1% (Acerbi and Tasche, 2002). In other words, $VaR_\varepsilon(x)$ of a portfolio return distribution is the ε-quantile of its distribution with negative sign:

$$VaR_\varepsilon(x) = -Q_\varepsilon(R_P(x)), \qquad (4.41)$$

where $Q_\varepsilon(R_P(x))$ is the ε-quantile of the portfolio return $R_P(x)$. The ε-quantile is equal to a real number r such that

$$Q_\varepsilon(R_P(x)) = F_{R_P}^{-1}(\varepsilon, x) = \inf\{r : F_{R_P}(r) > \varepsilon\},$$

where $F_{R_P}^{-1}$ is the inverse of the portfolio return cumulative distribution function. To support intuition, see Fig. 4.11. If the market returns have a multivariate normal distribution with zero means and covariance matrix Σ, then

$$VaR_\varepsilon(x) = \phi^{-1}(\varepsilon)\sigma_P(x),$$

where $\phi^{-1}(\varepsilon)$ is the $\varepsilon-$quantile of the standard normal distribution, and $\sigma_P^2(x) = x^T \Sigma x$.

Let $y_1, \cdots, y_n$ be the discrete values of a random variable, and let $\min^k\{y_1, \cdots, y_n\}$ denote the k-th smallest value among $y_1, \cdots, y_n$. The empirical ε-quantile of the sample $R_{P,1}(x), \ldots, R_{P,T}(x)$ is $Q_\varepsilon(x) = \min^{\lfloor \varepsilon T \rfloor + 1}\{R_{P,1}(x), \ldots, R_{P,T}(x)\}$, where $\lfloor \varepsilon T \rfloor$ is the largest integer not exceeding εT. Thus, we can write that

$$VaR_\varepsilon(x) = -\min^{\lfloor \varepsilon T \rfloor + 1}\{R_1(x), \ldots, R_T(x)\}$$
$$= \max^{\lfloor \varepsilon T \rfloor + 1}\{-R_1(x), \ldots, -R_T(x)\}. \qquad (4.42)$$

In the following exercise, we propose how to compute in practice the VaR of a portfolio for a fixed confidence level related to a predefined time horizon T.

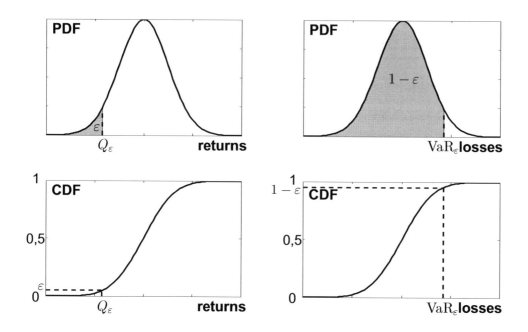

Figure 4.11: Example of VaR

The Mean-VaR model can be expressed by the following bi-objective optimiza-

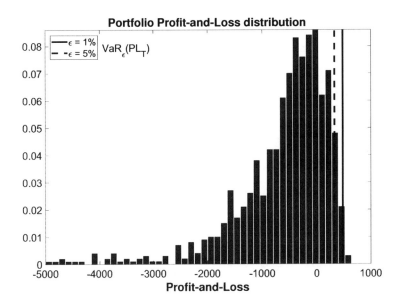

Figure 4.12: Example of VaR of portfolio Profit and Loss

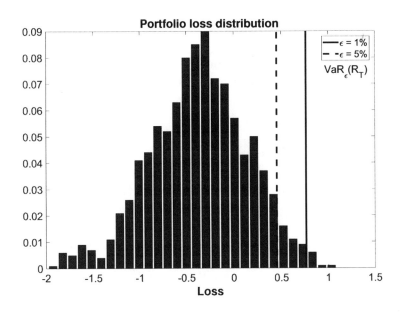

Figure 4.13: Example of VaR of portfolio losses

tion problem:

$$
\begin{cases}
\min & VaR_\varepsilon(x) \\
\max & \mu_P(x) = \sum_{i=1}^{n} \mu_i x_i \\
\text{s.t.} & \\
& u^T x = 1 \\
& x \geq 0
\end{cases}
\tag{4.43}
$$

As described by Benati and Rizzi (2007), the Mean-VaR model can be formulated as a mixed integer linear programming problem which is a non-convex optimization problem and has a computational burden higher than that of LP. The reason for the popularity of VaR as a risk measure is likely due to its interpretation as a representation of high losses. As mentioned above, when the returns are normally distributed, the task of computing VaR is reduced to computing the standard deviation of the portfolio. However, when the returns are not-normally distributed, VaR is difficult to optimize, because, in this case, it is a non-convex function, that may exhibit many local minima. Therefore, the VaR minimization is a combinatorial optimization, where its computational complexity is high.

4.2.8   Mean-Conditional Value-at-Risk model

The definition of the Conditional Value-at-Risk (CVaR, often called *average Value-at-Risk* or *expected shortfall*), at the specified confidence level ε, is the mathematical transcription of the concept of average of losses in the worst $100\varepsilon\%$ of cases (see Acerbi and Tasche, 2002). As said in the previous section, typical settings are $\varepsilon = 0.01, 0.05$ or 0.10, and as usual losses are defined as negative outcomes of returns. CVaR is approximatively the expected loss exceeding VaR (at the same confidence level ε). The equality is exact if the ε-quantile of the portfolio return $R_P(x)$ is unique, which is always true when $R_P(x)$ is a continuous random variable. CVaR is known to have better properties than VaR. This is due to theoretical and computational reasons. From the theoretical viewpoint, CVaR satisfies the axioms of a coherent risk measure (Artzner et al, 1999) and, furthermore, the mean-CVaR model is consistent with second-order stochastic dominance (Ogryczak and Ruszczynski, 2002). From the computational viewpoint, the mean-CVaR model can be efficiently solved by means of linear programming (Rockafellar and Uryasev, 2000).
Formally, if we denote by $L_P(x)$ the portfolio loss, we can define CVaR as:

$$
CVaR_\varepsilon(x) = E[L_P(x)|L_P(x) \geq VaR_\varepsilon(x)] .
\tag{4.44}
$$

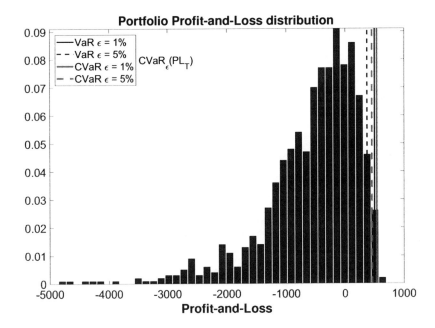

Figure 4.14: Example of CVaR of portfolio Profit and Loss

In the following exercise, we show how to compute in practice CVaR of a portfolio for a fixed confidence level ε related to a predefined time horizon T.

Exercise 138 (CVaR example) *Let us continue from Exercise 137. Load the workspace* PortPLandRet.mat; *then, using the Profit and Loss vector* PL_T *of the portfolio and the future portfolio return* R_T, *compute* $CVaR_\varepsilon(PL_T)$ *and* $CVaR_\varepsilon(R_T)$ *at confidence levels* $\varepsilon = [0.01, 0.05]$ *related to the predefined time horizon* $T = 1$ *month. Duplicate Figs. 4.14 and 4.15, where we report the empirical pdf of* $-PL_T$ *and of* $-R_T$, *respectively.*

Sol.: See Script S_CVaRExample.

In the case of continuous random variables, we can write

$$CVaR_\varepsilon(x) = \frac{1}{\varepsilon} \int\limits_{l(x,t) \geq VaR_\varepsilon(x)} l(x,t) p_{L_P(x)}(t) dt, \qquad (4.45)$$

where $VaR_\varepsilon(x) = -Q_\varepsilon(R_P(x))$ and $p_{L_P(x)}$ is the probability density function of the portfolio losses. Note that if $\varepsilon \to 0$, then $CVaR_\varepsilon(x)$ tends to the maximum loss of the portfolio. While, if $\varepsilon \to 1$, then $CVaR_\varepsilon(x)$ tends to the portfolio

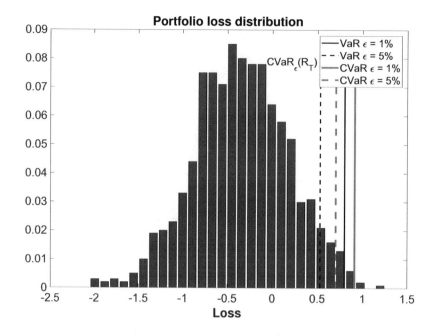

Figure 4.15: Example of CVaR of portfolio losses

expected loss, or, equivalently, $CVaR_{\varepsilon=1}(x) = -E[R_P(x)]$. Furthermore, we observe that

$$
\begin{aligned}
CVaR_\varepsilon(x) &= \frac{1}{\varepsilon} \int\limits_{l(x,t) \geq VaR_\varepsilon(x)} l(x,t) p_{L_P(x)}(t) dt \\
&\geq \frac{1}{\varepsilon} \int\limits_{l(x,t) \geq VaR_\varepsilon(x)} VaR_\varepsilon(x) p_{L_P(x)}(t) dt \\
&= VaR_\varepsilon(x) \frac{1}{\varepsilon} \int\limits_{l(x,t) \geq VaR_\varepsilon(x)} p_{L_P(x)}(t) dt \\
&= VaR_\varepsilon(x) \,,
\end{aligned}
$$

since $\displaystyle\int\limits_{l(x,t) \geq VaR_\varepsilon(x)} p_{L_P(x)}(t) dt = \varepsilon$. This implies that CVaR is always at least as large as VaR (see Figs. 4.14, 4.15, and 4.16), so that CVaR can be considered as an upper bound of VaR. Hence, minimizing CVaR of a portfolio is closely related to minimizing VaR. In addition, as shown in Fig. 4.16, we can note that

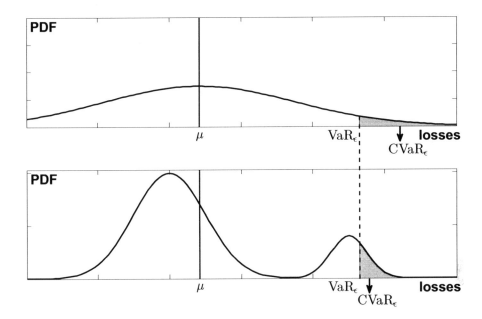

Figure 4.16: Example of VaR and CVaR for different probability density functions

for a fixed confidence level ε different random variables can have the same VaR but different CVaR, and vice versa.

In the case where there does not exist a value r such that $P(R_P(x) \leq r) = \varepsilon$, we can consider an upper ε-quantile

$$Q^{\varepsilon}(R_P(x)) = \inf \{r : F(r) = P(R_P(x) \leq r) > \varepsilon\},$$

and the excess probability $P(R_P(x) \leq Q^{\varepsilon}(R_P(x))) - \varepsilon$ is taken out from the conditional expectation of values of $R_P(x)$ below the ε-quantile. Hence, the formal definition of CVaR is the following:

$$
\begin{aligned}
CVaR_{\varepsilon} = -\frac{1}{\varepsilon} & E\left[R_P(x)\mathbb{1}_{\{R_P(x)\leq Q^{\varepsilon}(R_P(x))\}}\right] \\
& - Q^{\varepsilon}(R_P(x))[P(R_P(x) \leq Q^{\varepsilon}(R_P(x))) - \varepsilon],
\end{aligned} \tag{4.46}
$$

where $\mathbb{1}_{relation}$ is the indicator function that is equal to 1 if the *relation* is true, and 0 otherwise; $Q^{\varepsilon}(R_P(x))$ is the upper ε-quantile of the random variable portfolio return as defined above.

An alternative definition of CVaR is:

$$CVaR_\varepsilon(x) = -\frac{1}{\varepsilon} \int_0^\varepsilon Q_{R_P(x)}(\alpha)d\alpha\,,$$

where $Q_{R_P(x)}(\alpha)$ is the α-quantile function of the portfolio return $R_P(x)$.

From the computational viewpoint, to find $CVaR_\varepsilon(x)$, we can consider the following auxiliary function:

$$F_\varepsilon(x,\zeta) = \zeta + \frac{1}{\varepsilon} \int_{-\infty}^{+\infty} [l(x,t) - \zeta]^+ p_{L_P(x)}(t)dt\,, \qquad (4.47)$$

where $[g]^+ = \max\{g,0\}$, i.e., $[l(x,t) - \zeta]^+ = \max\{l(x,t) - \zeta, 0\}$. As shown by Rockafellar and Uryasev (2000), the Conditional Value-at-Risk of a portfolio x is equal to the minimum value on ζ of the auxiliary function (4.47), namely

$$CVaR_\varepsilon(x) = \min_{\zeta \in \mathbb{R}} F_\varepsilon(x,\zeta)\,. \qquad (4.48)$$

Furthermore, minimizing $CVaR_\varepsilon(x)$ w.r.t. x is equivalent to minimizing $F_\varepsilon(x,\zeta)$ w.r.t. x and ζ as follows:

$$\min_{x \in C} CVaR_\varepsilon(x) = \min_{x \in C} \min_{\zeta \in \mathbb{R}} F_\varepsilon(x,\zeta) = \min_{(x,\zeta) \in C \times \mathbb{R}} F_\varepsilon(x,\zeta), \qquad (4.49)$$

where C is the set of feasible portfolios and ζ is a scalar. In the discrete case, we can write Expression (4.47) as follows:

$$F_\varepsilon(x,\zeta) = \zeta + \frac{1}{\varepsilon} \sum_{t=1}^T p_t [l(x,t) - \zeta]^+, \qquad (4.50)$$

where, for $t = 1,\ldots,T$, p_t is the probability that the scenario $l(x,t)$ occurs. Furthermore, considering the losses as negative outcomes of portfolio return and assuming that all scenarios $l(x,t)$ are equally likely, we have $l(x,t) = -r_P(x,t) = -\sum_{i=1}^n r_{it}x_i$ and $p_t = 1/T$. Hence, Expression (4.50) becomes

$$F_\varepsilon(x,\zeta) = \zeta + \frac{1}{\varepsilon T} \sum_{t=1}^T \left[\sum_{i=1}^n -r_{it}x_i - \zeta \right]^+. \qquad (4.51)$$

Similar to the Mean-Risk models previously described, the Mean-CVaR model can be expressed by the following bi-objective optimization problem:

$$
\begin{cases}
\min & CVaR_\varepsilon(x) \\
\max & \mu_P(x) = \sum_{i=1}^{n} \mu_i x_i \\
\text{s.t.} & \\
& u^T x = 1 \\
& x \geq 0
\end{cases}
\tag{4.52}
$$

where μ is the vector of expected returns of the assets, u is the all-ones vector. Using the ε-constraint method (see Section 3.3.1), the multi-objective optimization problem (4.32) can be formulated as the following single-objective optimization problem

$$
\begin{cases}
\min & CVaR_\varepsilon(x) \\
\text{s.t.} & \\
& \mu^T x \geq \eta \\
& u^T x = 1 \\
& x \geq 0
\end{cases}
\quad \Rightarrow \quad
\begin{cases}
\min & CVaR_\varepsilon(x) \\
\text{s.t.} & \\
& \mu^T x = \eta \\
& u^T x = 1 \\
& x \geq 0
\end{cases}
\tag{4.53}
$$

where η is the required level of the portfolio expected return, and the implication is due to the convexity of Problem (4.53). Then, substituting (4.51) in (4.49), in the case of discrete random variables Problem (4.53) can be formulated as

$$
\begin{cases}
\min_{(x,\zeta)} & \zeta + \dfrac{1}{\varepsilon T} \sum_{t=1}^{T} \max(-\sum_{i=1}^{n} r_{it} x_i - \zeta, 0) \\
\text{s.t.} & \\
& \sum_{i=1}^{n} \mu_i x_i = \eta \\
& \sum_{i=1}^{n} x_i = 1 \\
& x_i \geq 0 \qquad\qquad\qquad\qquad\qquad i = 1, \dots, n
\end{cases}
\tag{4.54}
$$

Note that the objective function is non-linear. However, as in Rockafellar and Uryasev (2000), we can linearize Problem (4.54) by considering T auxiliary variables d_t defined as the deviation of the portfolio losses $-\sum_{i=1}^{n} r_{it} x_i$ from ζ when $-\sum_{i=1}^{n} r_{it} x_i > \zeta$, and 0 otherwise. Therefore, we substitute $\max(-\sum_{i=1}^{n} r_{it} x_i - \zeta, 0) = d_t$ by adding the following constraints: $d_t \geq 0$, $d_t \geq -\sum_{i=1}^{n} r_{it} x_i - \zeta$. Hence, the Mean-CVaR model can be rewritten as the following Linear Programming problem:

$$
\begin{cases}
\displaystyle \min_{(x,\zeta,d)} \quad \zeta + \frac{1}{\varepsilon}\frac{1}{T}\sum_{t=1}^{T} d_t \\[2mm]
\text{s.t.} \\[2mm]
\quad d_t \geq -\sum_{i=1}^{n} r_{it}x_i - \zeta \quad t = 1,\dots,T \\[2mm]
\quad d_t \geq 0 \qquad\qquad\qquad t = 1,\dots,T \\[2mm]
\quad \sum_{i=1}^{n} \mu_i x_i = \eta \\[2mm]
\quad \sum_{i=1}^{n} x_i = 1 \\[2mm]
\quad x_i \geq 0 \qquad\qquad\qquad i = 1,\dots,n \\[2mm]
\quad \zeta \in \mathbb{R}
\end{cases}
\tag{4.55}
$$

Note that when solving this optimization problem, under some assumptions, the optimal value of the variable ζ coincides with $VaR_\varepsilon(x^\star)$, where $x^\star$ is the optimal solution of Problem (4.55), or otherwise it can be considered as its approximation.

As shown by Rockafellar and Uryasev (2000), if the assets returns are normally distributed, then CVaR is proportional to volatility. Thus, the optimal portfolios obtained by the Mean-CVaR model and by the Markowitz model are the same. In the case of non-normal distribution of returns, these methods can reveal significant differences. It seems that, for low values of the confidence level ε, the discrepancy between the CVaR and the Markowitz solutions is higher. Indeed, it is important to point out that the portfolios of the Mean-CVaR efficient frontier strictly depend, by definition, on the left tail of the portfolio return distribution, which corresponds to high losses, while the Markowitz approach tends to penalize high losses as well as high gains. On the other hand, we highlight that the Mean-Variance model objective function is based on the correlation analysis among all assets.

Following the notation used to represent the generic LP problem (3.4), to solve Problem (4.55) by means of the built-in Function `linprog`, we can set its input parameters as follows:

$$
A = \begin{bmatrix}
-r_{1,1} & \cdots & -r_{n,1} & -1 & 0 & \cdots & 0 & -1 \\
-r_{1,2} & \cdots & -r_{n,2} & 0 & -1 & \cdots & 0 & -1 \\
\vdots & \ddots & \vdots & \vdots & \vdots & \ddots & \vdots & \vdots \\
-r_{1,T} & \cdots & -r_{n,T} & 0 & 0 & \cdots & -1 & -1
\end{bmatrix};
$$

158

$$b = \begin{pmatrix} 0 \\ 0 \\ \vdots \\ 0 \end{pmatrix} \; ; \qquad x = \begin{pmatrix} x_1 \\ \vdots \\ x_n \\ d_1 \\ \vdots \\ d_T \\ \zeta \end{pmatrix} \; ; \qquad f = \begin{pmatrix} 0 \\ \vdots \\ 0 \\ 1 \\ \frac{1}{\varepsilon T} \\ \vdots \\ 1 \\ \frac{1}{\varepsilon T} \\ 1 \end{pmatrix} \; ;$$

$$l_b = \begin{pmatrix} 0 \\ \vdots \\ 0 \\ 0 \\ \vdots \\ 0 \\ -\infty \end{pmatrix} \; ; \qquad u_b = \begin{pmatrix} +\infty \\ \vdots \\ +\infty \\ +\infty \\ \vdots \\ +\infty \\ +\infty \end{pmatrix} \; ;$$

$$b_{eq} = \begin{pmatrix} \eta \\ 1 \end{pmatrix} \; ; \qquad A_{eq} = \begin{pmatrix} \mu_1 & \cdots & \mu_n & 0 & \cdots & 0 & 0 \\ 1 & \cdots & 1 & 0 & \cdots & 0 & 0 \end{pmatrix}.$$

In the following exercise, we show how to determine the efficient frontier related to the Mean-CVaR model on a real market.

Exercise 139 (Mean-CVaR portfolios) *Write a script (named* S_CVaR*) that solves the following points.*

1. *Import data of* weekly_EUROSTOXX50_price_time *in a matrix* A. *Note that, in this file, there are weekly historical prices of the EuroStoxx 50 market index components. Define the* dates *vector picking only the first column of* A *and a matrix* P *selecting the remaining columns. Compute the linear returns from* P *creating a matrix* RR.

2. *Compute the assets expected returns vector* μ.

3. *Consider* $CVaR_\varepsilon$ *at the confidence levels* $\varepsilon = [0.01, 0.05]$ *as risk measure. Then, compute the return of the Minimum-Risk portfolio* (η_{min}) *and that of the Maximum-Return portfolio* (η_{max}) .

4. *Compute the* $1 \times N$ *vector* eta *of the target portfolio expected returns,*

159

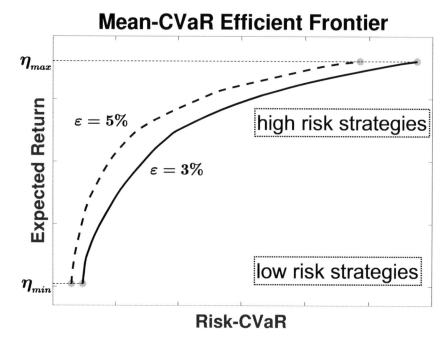

Figure 4.17: Example of the Mean-CVaR *efficient frontier* for different levels of target portfolio expected returns and for two different levels of ε

4.2.9 Mean-Gini model

The Gini's Mean Difference (often called Gini index) is a measure of dispersion that is widely used for evaluating income inequality. Yitzhaki (1982) proposed a Risk-Gain model, where risk is measured by the Gini's Mean Difference (GMD). This risk measure is the average of the absolute differences between all possible pairs of observations of a random variable. Thus, denoting by R' and R'' two identically distributed random variables, the Gini's Mean Difference is defined as

$$GMD = \frac{1}{2} E\left[\,|R' - R''|\,\right]. \tag{4.56}$$

Now, since from Remark 140 we have

$$
\begin{aligned}
|R' - R''| &= R' + R'' - 2\min\left(R', R''\right) \\
&= R' + R'' - 2\left(\min\left(0, R' - R''\right) + R''\right) \\
&= R' - R'' - 2\min\left(0, R' - R''\right),
\end{aligned}
$$

Expression (4.56) can be rewritten as follows

$$
\begin{aligned}
GMD &= \frac{1}{2}\left(E\left[R'\right] - E\left[R''\right] - 2E\left[\min\left(0, R' - R''\right)\right]\right) \\
&= -E\left[\min\left(0, R' - R''\right)\right] \tag{4.57} \\
&= E\left[\max\left(0, R' - R''\right)\right], \tag{4.58}
\end{aligned}
$$

where $E\left[R''\right] = E\left[R'\right]$, being R' and R'' identically distributed.

Remark 140 (Some relations of the min and max functions) *To obtain Expressions (4.57) and (4.58), we note that for $a, b \in \mathbb{R}$ these relations hold:*

- $\min(a, b) = \min(0, a - b) + b = \min(0, b - a) + a$;
- $\max(a, b) = \max(0, a - b) + b = \max(0, b - a) + a$;
- $a + b = \max(a, b) + \min(a, b)$;
- $|a - b| = \max(a, b) - \min(a, b)$;
- $|a - b| = a + b - 2\min(a, b) = -(a + b) + 2\max(a, b)$.

In the case of discrete random variables, the Gini's Mean Difference $GMD\,(x)$ corresponding to the portfolio random return $R(x)$ is

$$GMD\,(x) = \frac{1}{2} \sum_{t'=1}^{T} \sum_{t''=1}^{T} p_{t'} p_{t''} \left|R_{t'}(x) - R_{t''}(x)\right|, \tag{4.59}$$

161

where, for $t = 1, \ldots, T$, p_t is the probability that the scenario $R_t(x)$ occurs. Furthermore, assuming (as usual in portfolio selection) that each outcome of the portfolio return $R_t(x)$ is equally likely, i.e., $p_t = \dfrac{1}{T}$, we can write

$$
\begin{aligned}
GMD\,(x) \;&=\; \frac{1}{2}\frac{1}{T^2}\sum_{t'=1}^{T}\sum_{t''=1}^{T} |R_{t'}(x) - R_{t''}(x)| \\
&=\; \frac{1}{T^2}\sum_{t'=1}^{T}\sum_{t''=1}^{T} \mathbb{1}_{R_{t''}(x)<R_{t'}(x)}\,(R_{t'}(x) - R_{t''}(x)) \\
&=\; \frac{1}{T^2}\sum_{t'=1}^{T}\sum_{t''=1}^{T} \max\{0, R_{t'}(x) - R_{t''}(x)\}\,.
\end{aligned}
$$

The risk-return analysis based on GMD as risk measure can be obtained by solving the following bi-objective optimization problem:

$$
\begin{cases}
\min \quad GMD\,(x) = \dfrac{1}{T^2}\displaystyle\sum_{t'=1}^{T}\sum_{t''=1}^{T} \max\{0, R_{t'}(x) - R_{t''}(x)\} \\[2ex]
\max \quad \mu_P(x) = \displaystyle\sum_{i=1}^{n} \mu_i x_i \\[2ex]
\text{s.t.} \\
\qquad u^T x = 1 \\
\qquad x \geq 0
\end{cases}
\tag{4.60}
$$

where u is the all-ones vector. Using the ε-constraint method (see Section 3.3.1), the multi-objective optimization problem (4.32) can be formulated as the following single-objective optimization problem

$$
\begin{cases}
\min \quad GMD\,(x) \\
\text{s.t.} \\
\qquad \mu^T x \geq \eta \\
\qquad u^T x = 1 \\
\qquad x \geq 0
\end{cases}
\quad\Rightarrow\quad
\begin{cases}
\min \quad GMD\,(x) \\
\text{s.t.} \\
\qquad \mu^T x = \eta \\
\qquad u^T x = 1 \\
\qquad x \geq 0
\end{cases}
\tag{4.61}
$$

where μ is the vector of the assets expected returns, η is the required level of the portfolio expected return, and the implication is due to the convexity of

Problem (4.61). Thus, the Mean-GMD model can be expressed as follows:

$$
\left\{
\begin{array}{ll}
\min & \dfrac{1}{T^2} \displaystyle\sum_{t'=1}^{T} \sum_{t''=1}^{T} \max\{0, R_{t'}(x) - R_{t''}(x)\} \\[2ex]
\text{s.t.} & \\
& \mu^T x = \eta \\
& u^T x = 1 \\
& x \geq 0
\end{array}
\right.
\tag{4.62}
$$

Now, similar to what we did for the Mean-CVaR model (see Section 4.2.8), we can linearize Problem (4.62) by introducing T^2 auxiliary variables $d_{t't''}$ (with $t', t'' = 1, \dots, T$) defined as the deviation of the portfolio return at times t', $R_{t'}(x) = \sum_{i=1}^{n} r_{it'} x_i$ from $R_{t''}(x) = \sum_{i=1}^{n} r_{it''} x_i$ when $R_{t'}(x) > R_{t''}(x)$, and 0 otherwise. Therefore, we substitute $\max\{0, R_{t'}(x) - R_{t''}(x)\} = d_{t't''}$ by adding the following constraints: $d_{t't''} \geq 0$, $d_{t't''} \geq \sum_{i=1}^{n} (r_{it'} - r_{it''}) x_i$. Hence, the Mean-GMD model can be rewritten as the following Linear Programming (LP) problem:

$$
\left\{
\begin{array}{ll}
\min & \dfrac{1}{2T^2} \displaystyle\sum_{t'=1}^{T} \sum_{t''=1}^{T} d_{t't''} \\[2ex]
\text{s.t.} & \\
& d_{t't''} \geq \displaystyle\sum_{i=1}^{n} (r_{it'} - r_{it''}) x_i, \quad t', t'' = 1, \dots, T \\[2ex]
& d_{t't''} \geq 0, \qquad\qquad\qquad\quad t', t'' = 1, \dots, T \\
& \mu^T x = \eta \\
& u^T x = 1 \\
& x \geq 0
\end{array}
\right.
\tag{4.63}
$$

Even though Problem (4.63) is an LP, it has $n + T^2$ variables and $2T^2 + n + 2$ constraints. Therefore, solving such a problem can be prohibitive even considering a few hundred historical scenarios.

To solve Problem (4.63) by means of the built-in Function `linprog`, we can set its input parameters as follows:

$$
A = \begin{bmatrix}
r_{1,1} - r_{1,1} & \cdots & r_{1,n} - r_{1,n} & -1 & 0 & \cdots & 0 & \cdots & 0 \\
r_{2,1} - r_{1,1} & \cdots & r_{2,n} - r_{1,n} & 0 & -1 & \cdots & 0 & \cdots & 0 \\
\vdots & \ddots & \vdots & \vdots & \vdots & \ddots & \vdots & \ddots & \vdots \\
r_{T,1} - r_{1,1} & \cdots & r_{T,n} - r_{1,n} & \vdots & \vdots & \ddots & \vdots & \ddots & \vdots \\
r_{1,1} - r_{2,1} & \cdots & r_{1,n} - r_{2,n} & \vdots & \vdots & \ddots & \vdots & \ddots & \vdots \\
r_{2,1} - r_{2,1} & \cdots & r_{2,n} - r_{2,n} & \vdots & \vdots & \ddots & \vdots & \ddots & \vdots \\
\vdots & \ddots & \vdots & \vdots & \vdots & \ddots & \vdots & \ddots & \vdots \\
r_{T,1} - r_{2,1} & \cdots & r_{T,n} - r_{2,n} & 0 & 0 & \cdots & -1 & \cdots & 0 \\
\vdots & \ddots & \vdots & \vdots & \vdots & \ddots & \vdots & \ddots & \vdots \\
r_{1,1} - r_{T,1} & \cdots & r_{1,n} - r_{T,n} & \vdots & \vdots & \ddots & \vdots & \ddots & \vdots \\
r_{2,1} - r_{T,1} & \cdots & r_{2,n} - r_{T,n} & \vdots & \vdots & \ddots & \vdots & \ddots & \vdots \\
\vdots & \ddots & \vdots & \vdots & \vdots & \ddots & \vdots & \ddots & \vdots \\
r_{T,1} - r_{T,1} & \cdots & r_{T,n} - r_{T,n} & 0 & 0 & \cdots & 0 & \cdots & -1
\end{bmatrix} ; \quad (4.64)
$$

$$
b = \begin{pmatrix} 0 \\ 0 \\ \vdots \\ 0 \\ \vdots \\ 0 \end{pmatrix} ; \;
x = \begin{pmatrix} x_1 \\ \vdots \\ x_n \\ d_{1,1} \\ d_{1,2} \\ \vdots \\ d_{t',t''} \\ \vdots \\ d_{T,T} \end{pmatrix} ; \;
f = \begin{pmatrix} 0 \\ \vdots \\ 0 \\ 1 \\ 1 \\ \vdots \\ 1 \\ \vdots \\ 1 \end{pmatrix} ; \;
l_b = \begin{pmatrix} 0 \\ \vdots \\ 0 \\ 0 \\ 0 \\ \vdots \\ 0 \\ \vdots \\ 0 \end{pmatrix} ; \;
u_b = \begin{pmatrix} +\infty \\ \vdots \\ +\infty \\ +\infty \\ +\infty \\ \vdots \\ +\infty \\ \vdots \\ +\infty \end{pmatrix}
$$

$$
b_{eq} = \begin{pmatrix} \eta \\ 1 \end{pmatrix} ; \qquad
A_{eq} = \begin{pmatrix} \mu_1 & \cdots & \mu_n & 0 & 0 & \cdots & 0 & \cdots & 0 \\ 1 & \cdots & 1 & 0 & 0 & \cdots & 0 & \cdots & 0 \end{pmatrix} .
$$

In the following exercise, we show how to determine the efficient frontier related to the Mean-Gini model on a real market.

Mean-Gini Efficient Frontier

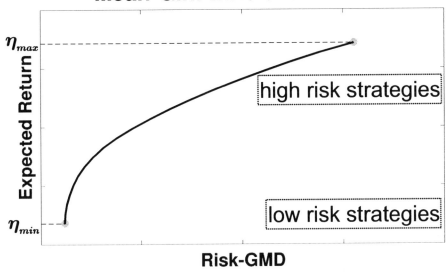

Figure 4.18: Example of Mean-GMD *efficient frontier* for different levels of target portfolio expected returns

---

Exercise 141 (Mean-Gini portfolios) *Write a Script that solves the following points.*

1. *Import data from* `weekly_EUROSTOXX50_price_time` *in a matrix* D, *choosing just the first 101 rows (for computational burden reasons). Then, define the* **dates** *vector considering only the first column of* D *and the matrix* P *obtained from the remaining columns of* D.

2. *Define a matrix* RR *that represents the linear returns obtained from* P.

3. *Compute the assets expected returns vector* μ.

4. *Define the matrix* A *with* $n + T^2 \times T^2$ *elements as in (4.64) used to represent the exponential number of inequalities constraint of Problem (4.63). To improve the efficiency of* `linprog`, *convert the full matrix* A *into the sparse matrix* A_tilde. *A sparse matrix is a matrix where many or most of the elements are zero. Then, if a matrix contains many zeros, converting it to sparse storage allows saving memory. In the Workspace check the memory (bytes) used to define* A *and* A_tilde. *Hint: see the built-in functions* `full` *and* `sparse`.

5. *Compute the return of the Minimum-Risk portfolio (η_{min}) and that of the Maximum-Return portfolio (η_{max}), then check the running time used to solve the LP problem (4.63) using the sparse matrix* A_tilde *and its full version* A. *Hint: see the built-in functions* tic *and* toc .*

6. *Compute the $1 \times N$ vector* eta *of the target portfolio expected returns, where $N = 20$. Note that $\eta \in [\eta_{min}, \eta_{max}]$.*

7. *For each value of* eta, *calculate the minimum GMD optimal portfolio and save the optimal risk values in the vector* Risk_GMD.

8. *Plot in the Return-Risk plane (* eta *vs* Risk_GMD*) the efficient frontier. Save the graph as* Gini_EF.jpg.

Sol.: See Script S_MeanGini.

4.3 Elements of bond portfolio immunization

In this section, we briefly describe how to hedge a portfolio of assets and liabilities in order to neutralize the effects of changes due to an interest rate fluctuation.

Let us consider, at time t, a positive cash flow (assets) $a = (a_1, a_2, ..., a_m)$ on the schedule $(t_1, t_2, ..., t_m)$, a negative cash flow (liabilities) $y = (y_1, y_2, ..., y_m)$ defined on the same schedule, and a term structure of interest rates $i(t, s)$, where $t \leq s$. In the classical approach to portfolio immunization, it is assumed that the changes in the term structure of interest rates are due to parallel shifts:

$$i\left(t^{+}, s\right) = i(t, s) + \Delta i .\tag{4.65}$$

where t^+ is a time immediately after t. To support intuition see Fig. 4.19. In the following remark we recall several indexes that are useful to introduce some classical models for portfolio immunization.

Remark 142 (Value of a cash flow, duration, convexity) *We denote the present value of a positive cash flow $a = (a_1, a_2, ..., a_m)$ on the schedule $(t_1, t_2, ..., t_m)$ by $V(t, a)$ where $t < t_1$, and in the case of a flat interest rate i we have*

$$V(t, a) = \sum_{k=1}^{m} a_k (1 + i)^{-(t_k - t)} .\tag{4.66}$$

Now, let us introduce the duration of an investment $a = (a_1, a_2, ..., a_m)$

$$D(t, a) = \frac{\sum\limits_{k=1}^{m} (t_k - t) a_k (1 + i)^{-(t_k - t)}}{V(t, a)} = \sum_{k=1}^{m} (t_k - t) p_k \tag{4.67}$$

166

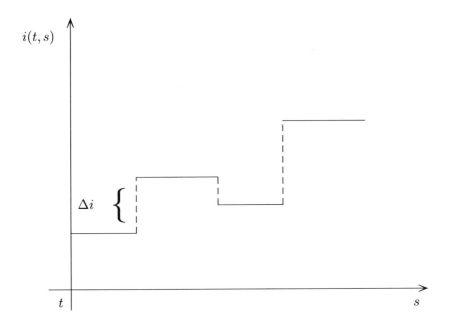

Figure 4.19: Example of parallel shifts in a constant term structure of interest rates

where $p_k = \dfrac{a_k \left(1+i\right)^{-(t_k-t)}}{V(t,a)}$. From a "physical" viewpoint, the duration can be seen as the temporal barycenter of the investment, represented by the weights p_k; while, from a "statistical" viewpoint, it can be interpreted as the weighted average of the payment dates of the cash flows. The duration of $a = (a_1, a_2, ..., a_m)$ is strictly linked to the price sensitivity of the investment defined as $\dfrac{V'(t,a)}{V(t,a)}$. Indeed, since the derivative of $V(t,a)$ w.r.t. the interest rate i is

$$V'(t,a) \quad = \quad \sum_{k=1}^{m} -(t_k - t)a_k \left(1+i\right)^{-(t_k-t)-1} \qquad (4.68)$$

$$= \quad -\frac{1}{(1+i)} \sum_{k=1}^{m} (t_k - t)a_k \left(1+i\right)^{-(t_k-t)} , \qquad (4.69)$$

we have that

$$\frac{V'(t,a)}{V(t,a)} = -\frac{1}{(1+i)} D(t,a) \qquad (4.70)$$

This measure of price sensitivity is often called semi-elasticity.

The second-order duration of $a = (a_1, a_2, ..., a_m)$ is defined as

$$D^{(2)}(t,a) = \frac{\sum_{k=1}^{m}(t_k - t)^2 a_k (1+i)^{-(t_k - t)}}{V(t,a)} = \sum_{k=1}^{m}(t_k - t)^2 p_k .$$ (4.71)

where $p_k = \dfrac{a_k (1+i)^{-(t_k - t)}}{V(t,a)}$. From a "physical" viewpoint, the second-order duration can be seen as the temporal moment of inertia of the investment; while, from a "statistical" viewpoint, it can be interpreted as the time dispersion of the cash flows.

Another useful index for portfolio immunization is convexity, that measures the curvature of the relationship of bond prices with interest rates. Formally, convexity is defined as $\dfrac{V''(t,a)}{V(t,a)}$. Now, since the second derivative of the bond price $V(t,a)$ w.r.t. the interest rate i is

$$\begin{aligned} V''(t,a) &= \frac{1}{(1+i)^2}\sum_{k=1}^{m}(t_k - t)^2 a_k (1+i)^{-(t_k - t)} \\ &+ \frac{1}{(1+i)^2}\sum_{k=1}^{m}(t_k - t) a_k (1+i)^{-(t_k - t)} \end{aligned}$$

we can write that

$$\frac{V''(t,a)}{V(t,a)} = \frac{1}{(1+i)^2}\left(D^{(2)}(t,a) + D(t,a)\right)$$ (4.72)

Thus, convexity of a cash flow is directly proportional to its duration and second-order duration, as defined in (4.67) and (4.71), respectively.

Given the term structure of interest rates $i(t,s)$ as in (4.65), the aim of the portfolio immunization is to neutralize the effects of changes of interest rates on a portfolio of assets and liabilities during the period $[t^+, t_1]$, where $t < t^+ < t_1$. Schematically, the conditions to obtain the immunization of an asset-liability portfolio are the following:

1. the net value, at time t, of the portfolio cash flow $V_{Net}(t, a-y) = 0$, where a is a positive cash flow (assets), y is a negative cash flow (liabilities), and

$$V_{Net}(t, a - y) = V(t,a) - V(t,y) .$$

In other words, at time t there has to be the financial equilibrium between positive and negative cash flows, namely

$$V(t,a) = V(t,y);$$ (4.73)

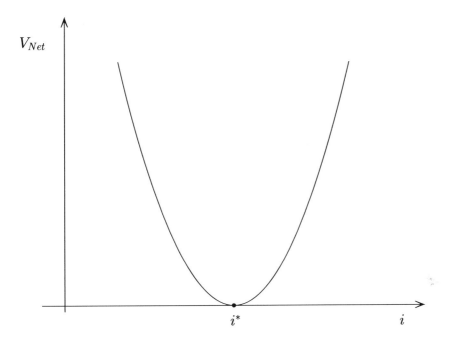

Figure 4.20: The portfolio net value as a function of the interest rate

2. at time t the portfolio net value $V_{Net}(t, a - y)$ has to be the lowest one
   with respect to changes of the interest rate i, namely $V_{Net}(t^+, a - y) \geq 0$,
   where t^+ is a time immediately after t. Thus, we should have

$$V\left(t^+, a\right) \geq V\left(t^+, y\right). \tag{4.74}$$

Condition (4.74) is intuitively illustrate in Fig. 4.20, in terms of the portfolio net value V_{Net} as a function of the interest rate i.

Sufficient conditions to pursue (4.74) are

$$\frac{dV_{Net}(t; i)}{di} = 0 \tag{4.75}$$

$$\frac{d^2 V_{Net}(t; i)}{di^2} \geq 0. \tag{4.76}$$

Condition (4.75) is equivalent to have

$$V'\left(t; a, i\right) = V'\left(t; y, i\right) \tag{4.77}$$

Now, using Relation (4.70) in (4.77) we can write

$$-\frac{1}{(1+i)}D(t, a)V(t, a) = -\frac{1}{(1+i)}D(t, y)V(t, y), \tag{4.78}$$

169

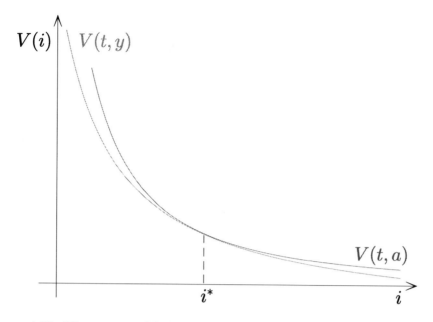

Figure 4.21: The asset and liability values as a function of the interest rate

and assuming that Condition (4.73) is satisfied, Condition (4.75) is therefore equivalent to have

$$D\left(a;i\right) = D\left(y;i\right) \tag{4.79}$$

where D is the duration of a cash flow, as in (4.67). Furthermore, imposing Condition (4.76) implies that

$$V''(t,a) \geq V''(t,y) \ . \tag{4.80}$$

Condition (4.80) is intuitively represented in Fig. 4.21. Also, using Expression (4.72) in (4.80), we have

$$\frac{1}{(1+i)^2}\left(D^{(2)}(t,a) + D(t,a)\right)V(t,a) \geq \frac{1}{(1+i)^2}\left(D^{(2)}(t,y) + D(t,y)\right)V(t,y).$$
$$\tag{4.81}$$

Hence, if Conditions (4.73) and (4.79) are satisfied, then Condition (4.76) is equivalent to

$$D^{(2)}\left(a;i\right) \geq D^{(2)}\left(y;i\right) \ , \tag{4.82}$$

where $D^{(2)}$ is the second-order duration of a cash flow, as in (4.71).
As proved by Fisher and Weil (1971), Conditions (4.73) and (4.79) ensure the immunization of an asset-liability portfolio with a *single* liability y both for infinitesimal and finite additive shifts of the interest rates. In the case of *multiple*

170

liabilities, Redington (1952) shows that an asset-liability portfolio is immunized w.r.t. changes of interest rates when Conditions (4.73), (4.79) and (4.82) are satisfied, but only for infinitesimal additive shifts. In the case of multiple liabilities and finite additive shift of term structure, we can substitute Condition (4.82) with the following Mean Absolute Deviation (MAD) constraints (see Fong and Vasicek, 1984; Shiu, 1986):

$$MAD(t_h, a) \geq MAD(t_h, y) \text{ for } h = 1, \ldots, m \qquad (4.83)$$

$$\sum_{j=1}^{n} x_j \sum_{k=1}^{m} |t_k - t_h| a_{ik} v(0, t_k) \geq \sum_{k=1}^{m} |t_k - t_h| y_k v(0, t_k) \text{ for } h = 1, \ldots, m.$$

In a nutshell, the MAD constraints (4.83) can be interpreted as the request that the cash flows of assets are more dispersed than those of liabilities. In Table 4.1 we list the classical immunization theorems which provide us with the conditions to neutralize the effects of the interest rates fluctuations in the procedures of selection of asset-liability portfolios.

Theorem	Liabilities	additive shifts	Immunization conditions
Fisher-Weil (Fisher and Weil, 1971)	single	finite	(4.73)-(4.79)
Rendington (Redington, 1952)	multiple	infinitasimal	(4.73)-(4.79)-(4.82)
General T. of Immunization (Fong and Vasicek, 1984) (Shiu, 1986)	multiple	finite	(4.73)-(4.79)-(4.83)

Table 4.1: Scheme of the classical approaches for the portfolio immunization.

Let us now consider a market with n bonds with cash flows described by an $m \times n$ matrix

$$A = \begin{bmatrix} a_{1,1} & a_{2,1} & \cdots & a_{n,1} \\ \vdots & \vdots & \ddots & \vdots \\ a_{1,m} & a_{2,m} & \cdots & a_{n,m} \end{bmatrix},$$

where each column represents the flow of the monetary amounts of a bond and each row represents the monetary amounts of the n bonds for each due time. Furthermore, let $x = (x_1, x_2, ..., x_n)$ be the shares to be allocated to each asset, and let $b = (b_1, b_2, ..., b_n)$ be the vector of bond prices at time $t = 0$, given by the market or computed by the following expression

$$b_j = \sum_{k=1}^{m} a_{jk} v\left(0, t_k\right) \quad \text{for} \quad j = 1, \ldots, n .$$

The value of the bond portfolio can be therefore defined as

$$V_P(x) = \sum_{j=1}^{n} x_j b_j .$$

From the modeling viewpoint, imposing Conditions (4.73) and (4.79) is equivalent to solving a system with $n - 2$ degree of freedom, where n is the number of decision variables x. Thus, to reduce the degree of freedom one can add further constraints or can include these conditions in an optimization problem. In this latter framework, for instance, we can seek the optimal shares vector x^* that satisfies the immunization conditions and minimizes the cost to purchase the bond portfolio as follows

$$\min V_P(x) = \sum_{j=1}^{n} x_j b_j .$$

An alternative request could be the maximization of the convexity of the assets A, namely

$$\max \sum_{j=1}^{n} x_j \sum_{k=1}^{m} t_k^2 a_{jk} v\left(0, t_k\right) .$$

Furthermore, one can also exclude the possibility of short selling, namely one can impose $x_j \geq 0$, with $j = 1, \ldots, n$.

For a portfolio of bonds, the immunization Conditions (4.73) and (4.79) can be respectively expressed as follows

$$\sum_{j=1}^{n} x_j V\left(0, a_j\right) = \sum_{k=1}^{m} y_k v\left(0, t_k\right) , \tag{4.84}$$

where $V\left(0, a_j\right) = \sum_{k=1}^{m} a_{jk} v\left(0, t_k\right)$, and

$$\frac{\sum_{j=1}^{n} x_j V\left(0, a_j\right) D\left(0, a_j\right)}{\sum_{j=1}^{n} x_j V\left(0, a_j\right)} = \frac{\sum_{k=1}^{m} t_k y_k v\left(0, t_k\right)}{\sum_{k=1}^{m} y_k v\left(0, t_k\right)} , \tag{4.85}$$

$$\text{where } D\left(0,a_j\right) = \frac{\sum\limits_{k=1}^{m} t_k a_{jk} v\left(0,t_k\right)}{V\left(0,a_j\right)}. \text{ Furthermore, using Relation (4.84) in (4.85)}$$

we obtain

$$\sum_{j=1}^{n} x_j V\left(0,a_j\right) D\left(0,a_j\right) = \sum_{k=1}^{m} t_k y_k v\left(0,t_k\right) . \qquad (4.86)$$

Thus, under the hypotheses of the General Theorem of Immunization, we can formulate the assets-liabilities portfolio selection model with minimum cost as the following Linear Programming problem

$$
\begin{cases}
\min\limits_{x} \ \sum\limits_{j=1}^{n} x_j b_j \\[2mm]
\text{s.t.} \\[1mm]
\quad \sum\limits_{j=1}^{n} x_j \sum\limits_{k=1}^{m} a_{jk} v\left(0,t_k\right) = \sum\limits_{k=1}^{m} y_k v\left(0,t_k\right) \\[2mm]
\quad \sum\limits_{j=1}^{n} x_j \sum\limits_{k=1}^{m} t_k a_{jk} v\left(0,t_k\right) = \sum\limits_{k=1}^{m} t_k y_k v\left(0,t_k\right) \\[2mm]
\quad \sum\limits_{j=1}^{n} x_j \sum\limits_{k=1}^{m} |t_k - t_h| x_{jk} v\left(0,t_k\right) \geq \sum\limits_{k=1}^{m} |t_k - t_h| y_k v\left(0,t_k\right) \quad h = 1,\dots,m \\[2mm]
\quad x_j \geq 0 \hspace{6cm} j = 1,\dots,n
\end{cases}
$$
$$(4.87)$$

In the following exercises, we show how to implement in practice Model (4.87).

Exercise 143 (Portfolio Immunization) *In the Script* S_Portfolio_Imm, *solve the following problem. Consider, at the evaluation date* $t = 17/9/2001$, *a cash flow with a time horizon of* 10 *years from* t, *and with a schedule having half-yearly payments, namely* $(t_1, t_2, ..., t_{20})$.
Then, load the file Data_Immunization.xls, *where there are stored:*

- *the term structure of prices into the sheet* TermStructPrices;

- *the cash flows of* 10 *bonds into the sheet* bonds;

- *the prices of the bonds at time* t *into the sheet* BondPrices.

The problem consists in selecting the immunized portfolio with minimum costs (under the finite additive shift hypothesis), in order to hedge a liability of 100€ *exigible in* $t_6 = 3$ *years (see the sheet* liability). *Solve the problem when no short selling is allowed.*

Sol.: See Script S_Portfolio_Imm.

Exercise 144 (Portfolio Immunization 2) *Given the same market conditions of Exercise 143, select the immunized portfolio with minimum costs (under the finite additive shift hypothesis), in order to hedge the negative cash flow available into the sheet* PortImm2 *of* Data_Immunization.xls.

Sol.: See Script S_Portfolio_Imm_2.

Exercise 145 (Portfolio Immunization 3) *Let us continue from Exercise 144. Select the immunized portfolio with minimum costs (under the finite additive shift hypothesis), in order to hedge a debt $D = 100€$ on a time horizon of 4 years with half-yearly payments. The debt is refunded by a French amortization (fixed payments, see Exercise 50) with an annual nominal interest rate $\bar{i} = 12\%$. Note that the negative cash flow is reported into the sheet* PortImm3 *of* Data_Immunization.xls *.*

Sol.: See Script S_Portfolio_Imm_3.

Part III

Derivatives pricing

Chapter 5

Further elements on Probability Theory and Statistics

In this chapter, we describe some tools which are commonly used for modeling stock prices and for pricing derivatives. In Section 5.1 we briefly introduce the Monte Carlo method, that is a numerical procedure generally used for solving complex problems. In Section 5.2 we give some basic concepts on stochastic processes, starting from the (discrete) random walk and its continuous time version, namely the Wiener process. Then, in Section 5.2.1 we discuss about the Brownian motion, which is generally used to represent the time evolution of the asset returns. Finally, Section 5.2.2 presents a simplified proof of Ito's Lemma, that we exploit to describe the geometric Brownian motion in Section 5.2.3.

For more on these topics, interested readers can refer to, e.g., Castellani et al (2006); Brandimarte (2013); Gnedenko (2018).

5.1 Introduction to Monte Carlo simulation

A simulation consists of investigating the behavior of a complex system by reproducing it in a controllable environment. An example of a simulation is the Monte Carlo method. It is particularly useful to reproduce and to numerically solve a problem that depends on many random variables and that is too complex. For instance, a simulation is able to test the effect of modifications of some input variables on those of output.

The main ingredients of the Monte Carlo method are:

- the input variables: for instance, we could assume that the input variable is a Student-t r.v., $X \sim StudT(0, 1, \nu)$ with $\nu = 5$. However, the input variables can be any random variables with any parameters.

- the output variables: variables that we suppose are linked to X by means of a model.

- the model: consider for simplicity the output $Y = \sin(X)$. The model is represented by one or more expressions that are functions of the input variable X and of some parameters. In a nutshell, the model provides the relations between inputs and outputs.

In practice, when a problem cannot be solved directly, e.g., it is not possible to analytically find a solution, it can be numerically addressed. Then, such a problem can be approached by producing a sufficiently high number of possible values which may be assumed by the input random variable. Accordingly, one can compute and analyze the corresponding outputs obtained by the model.

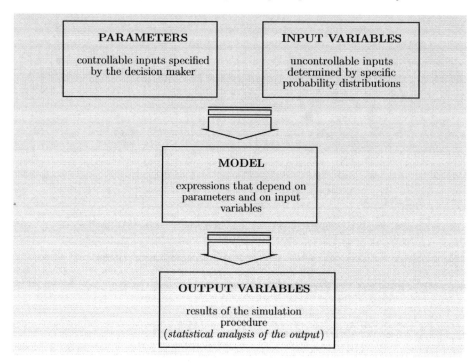

Figure 5.1: Scheme of the Monte Carlo simulation

The Monte Carlo method is based on the generation of random numbers[1] for the

---

[1]Specifically, we should talk about pseudo-random numbers (see, e.g., Brandimarte, 2013).

input variables, according to their joint probability distribution. These random numbers are fed into a simulation procedure, generating, in turn, a sequence of random numbers that represent a sample of the possible values which may be assumed by the output variables. Thus, using suitable statistical techniques, one can examine the estimation of this output and its possible errors. In Fig. 5.1, we report a synthetic scheme of the Monte Carlo simulation.

Exercise 146 (Monte Carlo method) *Setting* $n = 100000$, *solve the following items.*

1. *Generate n random numbers of the input variable Z, where Z is a standard normal r.v., i.e., $Z \sim N(0,1)$.*

2. *Plot the empirical pdf and the cdf of Z (see Ex. 99 and Ex. 95, respectively). Then, save these figures as* G_MCmethodIn_pdf.jpg *and* G_MCmethodIn_cdf.jpg.

3. *Given a "black box" (i.e., the model in Fig. 5.1)*

$$\phi(\cdot) = \frac{1}{2}\left[1 + \mathrm{erf}\left(\frac{(\cdot)}{\sqrt{2}}\right)\right],$$

   write a Function (named F_MCmethod*) which links the input variable Z to the output variable Y, namely $Y = \phi(Z)$.*

4. *Using the Function* F_MCmethod *and the Function in Ex. 100, write a Script* S_MCmethod *to compute the mean, the variance, the skewness and the kurtosis of Y.*

   Finally, plot the empirical pdf and cdf of Y, and save them as G_MCmethodOut_pdf.tif *and* G_MCmethodOut_cdf.tif, *respectively.*

5. *Think carefully about the output Y, and check if Y looks like a random variable described in Section 2.6.*

Sol.: See Function F_MCmethod and Script S_MCmethod.

5.2 Stochastic processes

The price of a stock at a future time t is unknown at the present, therefore it has to be modeled by a random variable. Furthermore, if we want to represent the

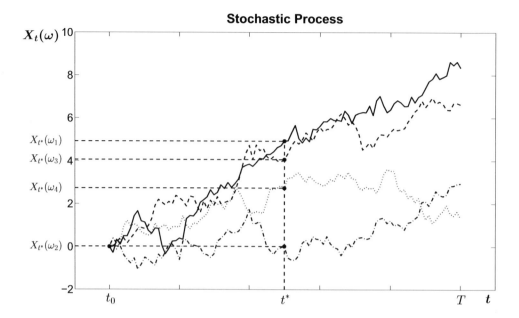

Figure 5.2: Example of a stochastic process

time evolution of the stock price, this is done considering a collection of random variables, i.e., by means of a stochastic process. As previously mentioned, in this section we provide some basic elements about stochastic processes and some useful tools to mathematically manipulate them.

Definition 147 *A stochastic process is a sequence $X_t(\omega)$ of random variables with $t \in (t_0, T]$ and $\omega \in \Omega$, where $X_{t_0} = x_0$ is known and Ω represents the set of possible states of Nature. In other words, when focusing on a specific time $t^* \in (t_0, T]$, then $X_{t^*}(\omega)$ is either a discrete or a continuous random variable. On the other hand, given an event $\omega^* \in \Omega$, the stochastic process $X_t(\omega^*)$ corresponds to a certain function of time, which is called the process path (or trajectory). To support intuition, see Fig. 5.2.*

Example 148 *Let t_0 be the current time and let P_t the price of an asset at time $t \geq t_0$, with $t \in [t_0, T]$. At a given state of nature ω^*, $P_t(\omega^*)$ corresponds to a specific evolution in time (or path) of the asset price. See Fig. 5.3.*

Now the question is how to model a stochastic process in finance. The rationale behind this issue is that, for instance, the price change (future price - current

180

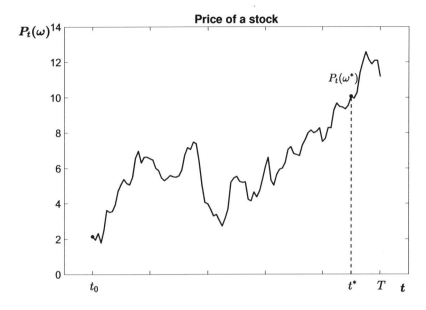

Figure 5.3: Example of a path of the asset price

price) is due to a deterministic component and to a random component. The first component models the low frequency dynamic (easily speaking, it could be interpreted as the average trend of the process), whereas the second part models the high frequency randomness of the process (to support intuition see Fig. 5.4). The fundamental unit for describing the random component is the Wiener process. However, to introduce the Wiener process, it is worth starting from a discrete random walk, which is the mathematical formalization of the idea of considering step-by-step increments in random directions.

Let us define the following collection of random variables, that represent the increments of a discrete random walk X (see Fig. 5.5) for each time step:

- $X_0 \equiv 0$

- $\Delta X_1 = \pm\sqrt{\Delta t}$ (with probability $\frac{1}{2}, \frac{1}{2}$)

- $\Delta X_2 = \pm\sqrt{\Delta t}$ (with probability $\frac{1}{2}, \frac{1}{2}$)

  $\vdots$

- $\Delta X_m = \pm\sqrt{\Delta t}$ (with probability $\frac{1}{2}, \frac{1}{2}$)

where ΔX_i and ΔX_j are independent and identically distributed (i.i.d.) random variables, with $\Delta X_i = X_i - X_{i-1}$.

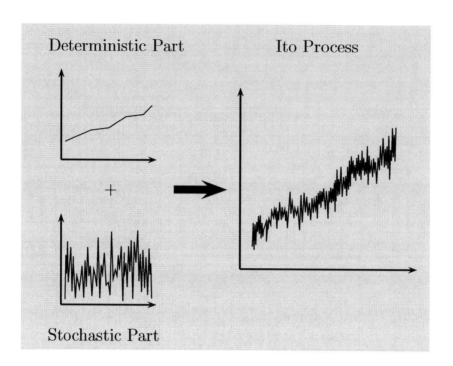

Figure 5.4: Intuitive scheme of an Ito process.

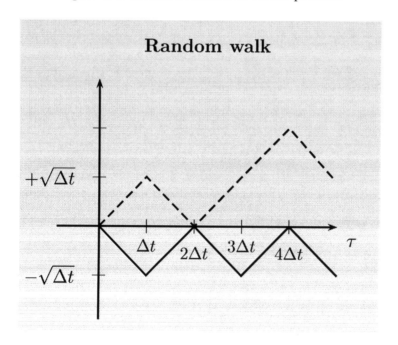

Figure 5.5: Example of a discrete random walk.

182

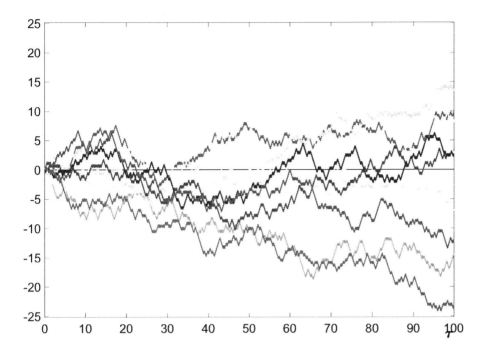

Figure 5.6: Example of paths of random walks.

Then, for all $i, j = 1, \ldots, m$ with $i \neq j$ we have

$$
\begin{aligned}
E\left[\Delta X_i\right] &= \frac{1}{2}\sqrt{\Delta t} - \frac{1}{2}\sqrt{\Delta t} = 0 \\
Var\left[\Delta X_i\right] &= E\left[\Delta X_i^2\right] = \Delta t \\
cov\left[\Delta X_i, \Delta X_j\right] &= E\left[\Delta X_i \Delta X_j\right] = 0 \,.
\end{aligned}
$$

Now, let us consider another collection of random variables, which represent the value of the random walk X (see Fig. 5.6) for each time step:

$$
\begin{aligned}
X_0 &\equiv 0 \\
X_1 &= X_0 + \Delta X_1 = \Delta X_1 \\
X_2 &= X_1 + \Delta X_2 = \Delta X_1 + \Delta X_2 \\
&\vdots \\
X_m &= X_{m-1} + \Delta X_m = \sum_{i=1}^{m} \Delta X_i
\end{aligned}
$$

183

where

$$E\left[X_m\right] \;=\; E\left[\sum_{i=1}^{m} \Delta X_i\right] = \sum_{i=1}^{m} E\left[\Delta X_i\right] = 0$$

and

$$Var\left[X_m\right] = E\left[X_m^2\right] - \underbrace{\left(E\left[X_m\right]\right)^2}_{=0}$$

$$= E\left[\left(\sum_{i=1}^{m} \Delta X_i\right)^2\right] = E\left[\sum_{i=1}^{m}\sum_{j=1}^{m} \Delta X_i \Delta X_j\right]$$

$$= \sum_{t=1}^{m} E\left[\Delta X_i^2\right] + \underbrace{\sum_{i\neq j} E\left[\Delta X_i \Delta X_j\right]}_{=0}$$

$$= \sum_{i=1}^{m} \Delta t = m\Delta t = \tau \; .$$

Furthermore, by the Central Limit theorem[2], we have that

$$X_m = \sum_{i=1}^{m} \Delta X_i \xrightarrow[m\to+\infty]{} N\left(0, \tau\right) \; . \tag{5.1}$$

Remark 149 *The Central Limit theorem states that the sum of m independent and identically distributed (i.i.d) random variables with finite variance tends to a normal distribution as the number m of random variable increases.*

The Wiener process can be defined as the continuous version of the discrete random walk, namely a random walk with infinitesimal time step dt as summarized in Table 5.1.

Definition 150 (Wiener process) *The Wiener process W_t with $t \in [t_0, T]$ is a continuous-time stochastic process characterised by the following properties:*

- *$W_{t_0} = 0$ (i.e., the process starts from zero);*

- *for any choice of $t, s \in [t_0, T]$, the increments $W_s - W_t$ are random variables independent and identically distributed (this theoretical condition holds often in financial applications, especially for $s - t \geq 15$ seconds);*

---

[2]see Lindeberg–Lévy theorem.

discrete time: random walk	continuous time: Wiener process
Δt (finite)	dt (infinitesimal)
$E[\Delta X_i] = 0$	$E[dW_t] = 0$
$Var\,[\Delta X_i] = \Delta t$	$Var[dW_t] = dt$
$cov\,[\Delta X_i, \Delta X_j] = E\,[\Delta X_i \Delta X_j] = 0$	$cov\,[dW_t, dW_{t'}] = E\,[dW_t dW_{t'}] = 0$
$\Delta X_i = \left\{+\sqrt{\Delta t}, \frac{1}{2}; -\sqrt{\Delta t}, \frac{1}{2}\right\}$	$dW_t \sim N(0, dt)$
ΔX_i i.i.d.	dW_t i.i.d.
$X_{i+1} = X_i + \Delta X_{i+1}$	$W_{t+dt} = W_t + dW_t$
$X_m = \sum_{i=1}^{m} \Delta X_i \xrightarrow[m \to +\infty]{} N(0, \tau)$	$W_\tau = \int_0^\tau dW_t \sim N(0, \tau)$

Table 5.1: Summary of the main features of the random walk and of the Wiener process.

- for any $t, s \in [t_0, T]$ with $t < s$, the random variable $W_s - W_t$ is normally distributed with mean equal to zero and variance equal to $s - t$. To see that intuitively, let us consider

$$W_s - W_t = \underbrace{dW_t + \ldots + dW_t}_{\text{where } dt + \ldots + dt = s - t},$$

then

$$
\begin{aligned}
E[W_s - W_t] &= E[dW_t + \ldots + dW_t] = 0; \\
Var[W_s - W_t] &= E[(W_s - W_t)^2] - \underbrace{E[(W_s - W_t)]^2}_{=0} \\
&= E[(dW_t + \ldots + dW_t)^2] \\
&= \underbrace{E[dW_t^2]}_{=dt} + \ldots + \underbrace{E[dW_t^2]}_{=dt} = s - t\,.
\end{aligned}
$$

Thus, similar to the discrete case (5.1), $W_s - W_t$ can be seen as the sum of i.i.d. random variables, that for the Central Limit theorem tends to a normal random variable, namely $W_s - W_t \sim N(0, s - t)$.

Now, let us explicitly define a Wiener process in terms of a standard normal random variable $Z \sim N(0, 1)$. For any $t, s \in [t_0, T]$ with $t < s$, we can write:

$$
\begin{aligned}
W_s - W_t &= K Z_t \\
E[W_s - W_t] &= E[K Z_t] = 0 \\
Var[W_s - W_t] &= E[(W_s - W_t)^2] \\
&= E[K^2 Z_t^2] = K^2 E[Z_t^2] = K^2
\end{aligned}
$$

Since $E[(W_s - W_t)^2] = s - t$, it implies that $K^2 = s - t \Rightarrow K = \sqrt{s - t}$. Thus

$$
W_s - W_t = Z_t \sqrt{s - t} \tag{5.2}
$$

In particular, if we set $t = t_0$, then

$$
\begin{aligned}
W_s &= Z_{t_0} \sqrt{s - t_0} \\
E[W_s] &= 0 \\
E[W_s^2] &= s - t_0
\end{aligned}
$$

To obtain the differential form of the Wiener process we can set $s = t + dt$ in (5.2), namely

$$
W_{t+dt} - W_t = dW_t
$$

where dW_t is the infinitesimal increment of the Wiener process. Thus,

$$
dW_t = Z_t \sqrt{dt} \quad .
$$

Remark 151 (Not differentiable paths) *Considering that*

$$
\frac{dW_t}{dt} = \lim_{\Delta t \to 0} \frac{\Delta W_t}{\Delta t} = \lim_{\Delta t \to 0} \frac{Z_t \sqrt{\Delta t}}{\Delta t} = \lim_{\Delta t \to 0} \frac{Z_t}{\sqrt{\Delta t}} = \pm \infty.
$$

This means that W_t is not differentiable at any point. To support intuition, see Fig. 5.7.

As described above, the Wiener process is the fundamental unit by which the random component of a stochastic process is modeled. In the next section we introduce the (arithmetic) Brownian motion, that is the natural generalization of the Wiener process, which is also called *standard* Brownian motion.

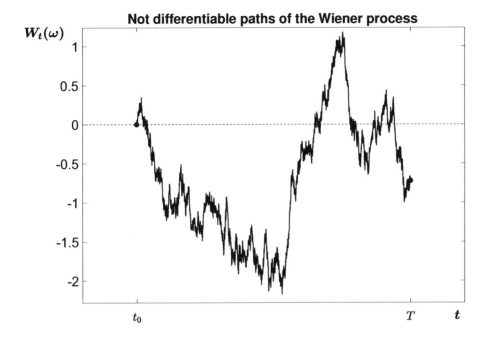

Figure 5.7: Example of a path of the Wiener process

5.2.1 Brownian motion

Starting from the Wiener process, we define the (arithmetic) Brownian motion as the stochastic process X_t, with $t \in [t_0, T]$ and $X_{t_0} = x_0$, that satisfies the following stochastic differential equation:

$$dX_t = \underbrace{\mu dt}_{\text{deterministic term}} + \underbrace{\sigma dW_t}_{\text{random term}}$$

$$\Downarrow$$

$$X_{t+dt} = X_t + dX_t = \underbrace{X_t + \mu dt}_{\text{known in } t} + \underbrace{\sigma Z_t \sqrt{dt}}_{\text{random in } t} \ ,$$

where μ and σ are called drift and diffusion coefficients, respectively. Let us consider some properties of the infinitesimal increment dX_t, namely its expected value and its variance:

- $E_t[dX_t] = E_t[\mu dt + \sigma dW_t] = \mu dt$;

- $E_t[(dX_t)^2] = E_t[\mu^2 dt^2 + 2\mu dt \sigma dW_t + \sigma^2 dW_t^2] = \mu^2 dt^2 + \sigma^2 dt$;

187

- $Var_t(dX_t) = E_t[(dX_t)^2] - E_t[dX_t]^2 = \mu^2 dt^2 + \sigma^2 dt - \mu^2 dt^2 = \sigma^2 dt.$

Note that E_t represents the expected value conditioned to the information available at time t.

Remark 152 *Consider an arithmetic Brownian motion*

$$dX_t = \mu dt + \sigma dW_t, \qquad (5.3)$$

where σ and μ are constant, $t \in [t_0, T]$, and $X_{t_0} = x_0$. We could introduce a new kind of integral, named stochastic, which has some properties of the primitive that are different from that of the ordinary integral (Riemann integral). Indeed, for example, unlike the ordinary integral, we have

$$\int_{t_0}^{T} X_t dX_t \neq \frac{1}{2} X_t^2 \Big|_{t_0}^{T}$$

as a consequence of Ito's lemma (see Section 5.2.2). However, for any stochastic process X_t, with $t \in [t_0, T]$, the following rule always holds:

$$\int_{t_0}^{T} dX_t = X_t \Big|_{t_0}^{T} = X_T - X_{t_0}. \qquad (5.4)$$

Exploiting the comments in Remark 152, let us consider the stochastic differential equation (5.3) where, for simplicity, $t_0 = 0$. This equation can be solved by integrating both the left and the right hand side as follows:

$$
\begin{aligned}
\int_0^T dX_t &= \int_0^T (\mu dt + \sigma dW_t) \\
X_T - X_0 &= \int_0^T \mu dt + \int_0^T \sigma dW_t \\
X_T - x_0 &= \mu T + \sigma(W_T - W_0) \\
X_T &= x_0 + \mu T + \sigma W_T,
\end{aligned}
$$

where $W_T \sim N(0, T)$ and the stochastic integrals are solved using the property (5.4). Thus, since

$$
\begin{aligned}
E_0[X_T] &= E_0[x_0 + \mu T + \sigma W_T] \\
&= E_0[x_0 + \mu T] + E_0[\sigma W_T] \\
&= x_0 + \mu T + \sigma \underbrace{E_0[W_T]}_{=0} = x_0 + \mu T
\end{aligned}
$$

188

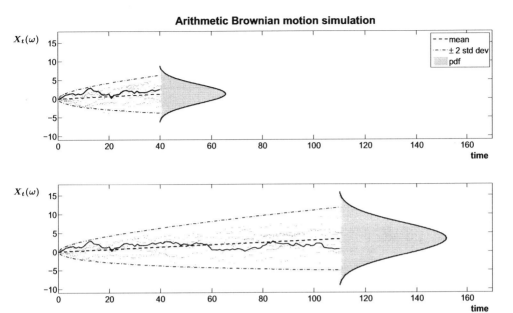

Figure 5.8: Evolution of an arithmetic Brownian motion

and

$$
\begin{aligned}
Var_0[X_T] &= E_0[(X_T - E_0[X_T])^2] \\
&= E_0[(X_T - x_0 - \mu T)^2] \\
&= E_0[(\sigma W_T)^2] = \sigma^2 E_0[W_T^2] = \sigma^2 T,
\end{aligned}
$$

we have that $X_T \sim N(x_0 + \mu T, \sigma^2 T)$. Hence, the pdf of X_T is

$$
f_{X_T}(u) = \frac{1}{\sqrt{2\pi\sigma^2 T}} e^{-\frac{(u - x_0 - \mu T)^2}{2\sigma^2 T}} \tag{5.5}
$$

Remark 153 (Numerical simulation schemes) *To simulate the trajectories of a stochastic process defined by a stochastic differential equation, the idea is to consider finite differences instead of differentials. The easiest discretization method is represented by the Euler scheme that, applied to eq. (5.3), becomes:*

$$
dX_t \approx \Delta X_t = \mu_t \Delta t + \sigma_t \Delta W_t = \mu_t \Delta t + \sigma_t \sqrt{\Delta t} Z_t. \tag{5.6}
$$

where $Z_t \sim N(0,1)$. However, there exist more accurate discretization methods such as the Milnstein scheme:

$$
\Delta X_t = \mu_t \Delta t + \sigma_t \Delta W_t + \frac{1}{2} \sigma_t \frac{\partial \sigma_t}{\partial X_t} \Delta t \left(Z_t^2 - 1\right). \tag{5.7}
$$

Exercise 154 (Numerical simulation of a Brownian motion)

Consider a Brownian motion X_t with drift μ and diffusion coefficient σ. Given a discrete time with a small time step Δt and using the Euler scheme introduced in Remark 153, the Brownian motion can be expressed by the following finite differences equation

$$\Delta X_k = \mu \Delta t + \sigma \Delta W_k \quad with \quad k = 1, \cdots, m,$$

where ΔW_k is the kth increment of the Wiener process (also called standard Brownian motion), namely a normally distributed random variable with mean 0 and standard deviation $\sqrt{\Delta t}$. More precisely, ΔW_k can be re-written as follows

$$\Delta W_k = \sqrt{\Delta t} Z_k, \tag{5.8}$$

where Z_k is a standard normal random variable.

Write a Function F_BrownMot to simulate the Brownian motion, considering the following steps.

1. *Let Z be an $n \times m$ matrix of random numbers generated from a standard normal distribution, where n is the number of simulations and m is the number of the time steps. Z can be obtained using* randn, *that approximately generates m independent standard normal random variables.*

2. *Compute the $n \times m$ matrix ΔX of the Brownian motion increments, considering that the random variable Z_k in (5.8) is the kth column of Z.*

3. *Determine the numerical simulation of the Brownian motion as follows*

$$X_k = X_0 + \sum_{j=1}^{k} \Delta X_j \quad for \quad k = 1, \ldots, m$$

using the built-in function cumsum, *and considering the starting point, if not provided, $X_0 = 0$ by default.*

Finally, using the function F_BrownMot, write a Script to solve the following points:

1. *simulate a Brownian motion with $n = 10000$, $m = 200$, $\mu = 0.1$, $\sigma = 0.4$ and $\Delta t = 0.1$.*

2. *Then, save the results in a .txt file.*

3. *Make the graph of the trajectories of X_t with $t \in [0, 20]$.*

4. *Plot the empirical pdf of the stochastic process for fixed times $t_k = k\Delta t$ with $k = 10, 30, ..., 170, 190$, to show the evolution of the random variable X_k. Then, save these figures as .jpg files.*

   Hint: use the Function F_Empirical_pdf *of Ex. 99.*

Sol.: See Function F_BrownMot and Script S_BrownMot.

Remark 155 *Note that X_T can assume negative values, in which case it cannot be used to model the price S_T of an asset, because $S_T \geq 0$. However, there exists a variable related to the value of an asset that can assume negative values: the absolute return $S_T - S_0$, and the relative return $\dfrac{S_T - S_0}{S_0}$ (see Fig. 5.9).*

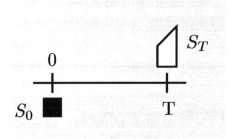

Figure 5.9: Scheme of the evolution of an equity price

- *$S_T - S_0$ is the absolute return between 0 and T and it can be viewed as an interest.*

- *$\dfrac{S_T - S_0}{S_0}$ is the return between 0 and T and it is similar to the interest rate.*

Following the rationale behind Remark 155, the idea is to model the infinitesimal relative return of an asset by means of the *arithmetic* Brownian motion:

$$dX_t = \frac{S_{t+dt} - S_t}{S_t}$$

$$\Rightarrow \mu dt + \sigma dW_t = \frac{dS_t}{S_t} \quad \text{with} \quad t \in [0, T] .$$

191

This means that the asset price S_t follows a *geometric* Brownian motion (see Section 5.2.3) and satisfies the following stochastic differential equation

$$dS_t \;=\; \mu S_t dt + \sigma S_t dW_t \;, \tag{5.9}$$

where $t \in [0, T]$ and $S_0 = s_0$ is known at time $t = 0$. In order to determine S_T, we could integrate both the left and the right hand side of (5.9) as follows:

$$
\begin{aligned}
\int_0^T \frac{dS_t}{S_t} &= \int_0^T \mu dt + \int_0^T \sigma dW_t \\
&= \mu T + \sigma \left(W_T - W_0 \right) \;,
\end{aligned}
$$

but, unlike the ordinary integral, we have $\int_0^T \frac{dS_t}{S_t} \neq \ln S_t |_0^T$ or, equivalently, $\frac{dS_t}{S_t} \neq d \ln S_t$. As mentioned in Remark 152, this is a consequence of Ito's lemma, that we discuss in the next section.

5.2.2   Ito's Lemma

Let us consider a function $F(X_t, t)$ of time t and of the stochastic process X_t, where $dX_t = \mu \left(X_t, t \right) dt + \sigma \left(X_t, t \right) dW_t$, $t \in [t_0, T]$, and $X_{t_0} = x_0$. For convenience, often we will synthetically indicate $\mu(X_t, t) = \mu_t$ and $\sigma(X_t, t) = \sigma_t$.

Remark 156 (Ito process)
The stochastic process $dX_t = \mu \left(X_t, t \right) dt + \sigma \left(X_t, t \right) dW_t$ is called Ito process, namely a process adapted in t that can be expressed by the sum of a deterministic and a random part. We assume that μ_t and σ_t are known given the information in t. Therefore, the process is said to be adapted in t. In words, it is a process that cannot "see into the future". Thus, an informal interpretation is that X_t is adapted in t if and only if, for every realization and every t, X_t is known at time t.

Then, the following result holds.

Lemma 157 (Ito) *Let $F(X_t, t)$ be twice differentiable in X_t and once in t, where $dX_t = \mu \left(X_t, t \right) dt + \sigma \left(X_t, t \right) dW_t$ is an Ito process, then*

$$dF(X_t, t) = \underbrace{\left[\frac{\partial F}{\partial t} + \mu_t \frac{\partial F}{\partial X_t} + \frac{1}{2} \sigma_t^2 \frac{\partial^2 F}{\partial X_t^2} \right]}_{\mu_F(X_t, t)} dt + \underbrace{\sigma_t \frac{\partial F}{\partial X_t}}_{\sigma_F(X_t, t)} dW_t \;. \tag{5.10}$$

Note that $dF(X_t, t) = \mu_F(X_t, t)dt + \sigma_F(X_t, t)dW_t$ is still an Ito process with μ_F and σ_F known, given the information in t.

Since a formal proof of Ito's Lemma contains a number of mathematical technical details that is beyond the scope of this book, we only present an informal justification of the result (5.10). This derivation is essentially based on the Taylor expansion of functions and on appropriate considerations on the infinitesimal increment of a Wiener process.

In the following remark, we briefly recall some concepts on the Taylor approximation of functions of one or more variables.

Remark 158 (Taylor approximation) *Let $f(x)$ be a function of one variable. We can obtain the differential $df(x) = f'(x)dx$ by the following 2 steps:*

1. *approximate $f(x + dx)$ by its Taylor expansion:*

$$f(x + dx) = f(x) + f'(x)dx + \frac{1}{2}f''(x)dx^2 + \ldots$$

2. *compute $df(x) = f(x + dx) - f(x)$, and neglect infinitesimals of order higher than 1:*

$$df(x) = f(x + dx) - f(x) = f'(x)dx + \underbrace{\frac{1}{2}f''(x)dx^2 + \ldots}_{\text{negligible}}$$

$$\Rightarrow df(x) \simeq f'(x)dx$$

Now, let $f(\mathbf{x})$ be a function of two or more variables. In general, the Taylor expansion of $f(\mathbf{x} + \mathbf{dx})$ can be written as follows:

$$f(\mathbf{x} + \mathbf{dx}) = f(\mathbf{x}) + \nabla f(\mathbf{x}) \cdot \mathbf{dx} + \frac{1}{2}\mathbf{dx}'\nabla^2 f(\mathbf{x})\mathbf{dx} + \ldots$$

where $\mathbf{x}$ is an $n \times 1$ vector of variables, $\nabla f(\mathbf{x})$ is the gradient of f, and $\nabla^2 f(\mathbf{x})$ is the Hessian matrix $(n \times n)$ of f.

In the case of a function $f(x, t)$ of two variables we can obtain the differential $df(x, t)$ by means of the following 2 steps, similarly to the case of a function of one variable:

1. *approximate $f(x + dx, t + dt)$ by its Taylor expansion:*

$$f(x+dx, t+dt) = f(x, t) + \frac{\partial f}{\partial t}dt + \frac{\partial f}{\partial x}dx + \frac{1}{2}\frac{\partial^2 f}{\partial t^2}dt^2 + \frac{1}{2}\frac{\partial^2 f}{\partial x^2}dx^2 + \frac{\partial^2 f}{\partial x \partial t}dxdt + \ldots$$

2. *compute $df(x,t) = f(x+dx, t+dt) - f(x,t)$, and neglect infinitesimals of order higher than 1:*

$$df(x,t) = \frac{\partial f}{\partial t}dt + \frac{\partial f}{\partial x}dx$$

$$+ \underbrace{\frac{1}{2}\frac{\partial^2 f}{\partial t^2}dt^2 + \frac{1}{2}\frac{\partial^2 f}{\partial x^2}dx^2 + \frac{\partial^2 f}{\partial x \partial t}dxdt + \dots}_{\text{negligible}}$$

$$\Rightarrow df(x,t) = \frac{\partial f}{\partial t}dt + \frac{\partial f}{\partial x}dx$$

Informal proof of Ito's Lemma. As described in Remark 158, since $F(X_t, t)$ is a function of the stochastic process X_t and of t, we can write

$$F(X_t + dX_t, t+dt) = F(X_t, t) + \frac{\partial F}{\partial t}dt + \frac{\partial F}{\partial X_t}dX_t$$

(5.11)

$$+ \frac{1}{2}\frac{\partial^2 F}{\partial t^2}dt^2 + \frac{1}{2}\frac{\partial^2 F}{\partial X_t^2}dX_t^2 + \frac{\partial^2 F}{\partial X_t \partial t}dX_t dt + \dots$$

Furthermore, we have

$$dX_t = \mu_t dt + \sigma_t dW_t \qquad (5.12)$$

where $dW_t = \sqrt{dt}Z_t$, with $Z_t \sim N(0,1)$. The rationale behind the informal justification of Ito's Lemma is to examine each addend of the Taylor expansion (5.11), to neglect all terms of order greater than dt, and, therefore, to keep all terms of order less than or equal to dt. Having said that, we observe that the first addend of (5.12) is proportional to dt, while the second addend is proportional to $\sqrt{dt}$, therefore both terms are not negligible. Now, consider the square of dX_t:

$$dX_t^2 = \underbrace{\mu_t^2 dt^2}_{\propto dt^2 (\text{negligible})} + \underbrace{2\mu_t\sigma_t dt dW_t}_{\propto dt^{\frac{3}{2}} (\text{negligible})} + \underbrace{\sigma_t^2 dW_t^2}_{\propto dt (\text{no negligible})} \qquad (5.13)$$

On the other hand,

$$dX_t dt = \underbrace{\mu_t dt^2}_{\propto dt^2 (\text{negligible})} + \underbrace{\sigma_t dt dW_t}_{\propto dt^{\frac{3}{2}} (\text{negligible})} \qquad (5.14)$$

Substituting (5.12),(5.13) and (5.14) in (5.11), we obtain that

$$F(X_t + dX_t, t+dt) = F(X_t, t) + \frac{\partial F}{\partial t}dt + \frac{\partial F}{\partial X_t}(\mu_t dt + \sigma_t dW_t) +$$

$$+ \frac{1}{2}\frac{\partial^2 F}{\partial X_t^2}(\sigma_t^2 dW_t^2) + \dots$$

194

Thus, we can approximate the differential $dF(X_t, t)$ as follows

$$dF(X_t, t) = \frac{\partial F}{\partial t} dt + \frac{\partial F}{\partial X_t} (\mu_t dt + \sigma_t dW_t) + \frac{1}{2} \frac{\partial^2 F}{\partial X_t^2} \sigma_t^2 dW_t^2 . \quad (5.15)$$

In order to complete the informal proof of Ito's Lemma, in the following remark, we discuss some properties of the quadratic variation of a Wiener process dW_t^2.

Remark 159 (Quadratic variation of a Wiener process) *Note that*

$$dW_t = Z_t \sqrt{dt} \Rightarrow dW_t^2 = Z_t^2 dt . \quad (5.16)$$

Since $Z_t \sim N(0,1)$, Z_t^2 is a chi-square random variable with 1 degree of freedom, $Z_t^2 \sim \chi^2_{\nu=1}$ (see Section 2.6.4) with

$$E[Z_t^2] = \nu = 1 \quad (5.17)$$
$$Var[Z_t^2] = 2\nu = 2 . \quad (5.18)$$

Thus, considering the expected value and the variance of the square of the Wiener process increment dW_t^2, we have

$$E[dW_t^2] = E[Z_t^2 dt] = E[Z_t^2] dt = dt$$
$$Var[dW_t^2] = Var[Z_t^2 dt] = Var[Z_t^2] dt^2 = 2 dt^2 .$$

This implies that the variance is an infinitesimal of order higher than dt, therefore dW_t^2 can be considered non-random, namely

$$dW_t^2 \simeq E[dW_t^2] = dt. \quad (5.19)$$

Using (5.19), Expression (5.15) can be reformulated, by conveniently rearranging the deterministic and the random terms, as follows

$$\begin{aligned} dF(X_t, t) &= \frac{\partial F}{\partial t} dt + \frac{\partial F}{\partial X_t} (\mu_t dt + \sigma_t dW_t) + \frac{1}{2} \frac{\partial^2 F}{\partial X_t^2} \sigma_t^2 dt \\ &= \left[\frac{\partial F}{\partial t} + \frac{\partial F}{\partial X_t} \mu_t + \frac{1}{2} \frac{\partial^2 F}{\partial X_t^2} \sigma_t^2 \right] dt + \frac{\partial F}{\partial X_t} \sigma_t dW_t . \quad (5.20) \end{aligned}$$

Example 160 *Compute the differential of $F(X_t, t)$, where $F(X_t, t) = \frac{1}{2} W_t^2$ and $dX_t = \mu_t dt + \sigma_t dW_t$ with $\mu_t = 0$ and $\sigma_t = 1$. Then, show that $\int_0^T W_t dW_t \neq \frac{1}{2} W_t^2 \big|_0^T$.*

195

Given $F(X_t, t) = \frac{1}{2}W_t^2$ we have

$$\frac{\partial F}{\partial t} = 0 ; \quad \frac{\partial F}{\partial W_t} = W_t ; \quad \frac{\partial^2 F}{\partial W_t^2} = 1 . \tag{5.21}$$

Thus, substituting (5.21) in (5.20), we obtain that

$$dF(W_t, t) = \left[0 + W_t \cdot 0 + \frac{1}{2} \cdot 1 \cdot 1^2 \right] dt + W_t \cdot 1 \cdot dW_t$$

$$\Rightarrow \quad d\left(\frac{1}{2}W_t^2 \right) = \frac{1}{2}dt + W_t dW_t$$

$$\Rightarrow \quad W_t dW_t = d\left(\frac{1}{2}W_t^2 \right) - \frac{1}{2}dt$$

$$\Rightarrow \quad \int_0^T W_t dW_t = \frac{1}{2}W_t^2 \Big|_0^T - \frac{1}{2}t \Big|_0^T$$

$$= \frac{1}{2}(W_T^2 - \underbrace{W_0^2}_{=0}) - \frac{1}{2}T .$$

5.2.3   Geometric Brownian motion

As mentioned at the end of Section 5.2.1, we model the asset price S_t by means of a *geometric* Brownian motion, namely a stochastic process that satisfies the following stochastic differential equation:

$$dS_t = \mu S_t dt + \sigma S_t dW_t \quad \text{with} \quad t \in [0, T] \quad \text{and} \quad S_0 = s_0 . \tag{5.22}$$

To find the analytic solution to (5.22), we now consider the function $F(S_t, t) = \ln S_t$, where $dS_t = \mu S_t dt + \sigma S_t dW_t$, and we compute its differential using Ito's Lemma. For the geometric Brownian motion we have

$$\mu_t = \mu S_t ; \qquad \sigma_t = \sigma S_t ;$$

$$\frac{\partial F}{\partial t} = 0 ; \qquad \frac{\partial F}{\partial S_t} = \frac{1}{S_t} ; \qquad \frac{\partial^2 F}{\partial S_t^2} = -\frac{1}{S_t^2} . \tag{5.23}$$

Thus substituting (5.23) in (5.20) we obtain

$$
\begin{aligned}
d(\ln S_t) &= \left(0 + \frac{1}{S_t}\mu_t - \frac{1}{2}\frac{1}{S_t^2}\sigma_t^2 \right) dt + \frac{1}{S_t}\sigma_t dW_t \\
&= \left(\frac{1}{S_t}\mu S_t - \frac{1}{2}\frac{1}{S_t^2}\sigma^2 S_t^2 \right) dt + \frac{1}{S_t}\sigma S_t dW_t \\
&= \left(\mu - \frac{1}{2}\sigma^2 \right) dt + \sigma dW_t \neq \frac{dS_t}{S_t} .
\end{aligned}
\tag{5.24}
$$

Now, let us compute S_T by integrating both sides of (5.24) over the interval $[0, T]$:

$$\int_0^T d(\ln S_t) = \int_0^T \left(\mu - \frac{1}{2}\sigma^2\right) dt + \int_0^T \sigma dW_t$$

$$\ln S_T - \ln s_0 = \left(\mu - \frac{1}{2}\sigma^2\right) T + \sigma(W_T - W_0)$$

$$\ln S_T = \ln s_0 + \left(\mu - \frac{1}{2}\sigma^2\right) T + \sigma W_T. \tag{5.25}$$

Note that $\ln S_T$ is normally distributed with mean $\ln s_0 + \left(\mu - \frac{1}{2}\sigma^2\right) T$ and variance $\sigma^2 T$. Thus, S_T is distributed as a log-normal random variable (see Section 2.6.3), namely

$$S_T = s_0\, e^{\left(\mu - \frac{1}{2}\sigma^2\right)T + \sigma W_T} \tag{5.26}$$

$$= s_0\, e^{\left(\mu - \frac{1}{2}\sigma^2\right)T + \sigma\sqrt{T}\,Z_T} \tag{5.27}$$

where $Z_T \sim N(0,1)$.

Exercise 161 (Equity price simulation) *Write a Function* F_PriceSim *that simulates* n *values of the price* S_T *of an asset at a generic date* T. *Under the hypothesis of the Black-Scholes model (see Section 6.2), we have*

$$S_T = S_0 e^{\left(\mu - \frac{1}{2}\sigma^2\right)T + \sigma\sqrt{T} Z_T},$$

where S_0 *is the current price of the asset,* Z_T *is a random variable generated from a standard normal distribution, and* μ *and* σ *are the drift and diffusion coefficients, respectively.*

Thus using the Function F_PriceSim, *write a Script that allows for the simulation of the price of the asset with* $S_0 = 100$, $\sigma = 0.2$, $\mu = 0.1$, $T = 1$, *and* $n = 10000$. *Furthermore, plot the pdf of* S_T *using the Function* F_Empirical_pdf *of Ex. 99.*

Sol.: See Function F_PriceSim and Script S_PriceSim.

Exercise 162 (Numerical simulation of a GBM) *Write a Script that simulates the paths of a Geometric Brownian Motion (GBM),*

$$S_{t_k} = S_0 e^{\left(\mu - \frac{1}{2}\sigma^2\right)t_k + \sigma W_{t_k}},$$

197

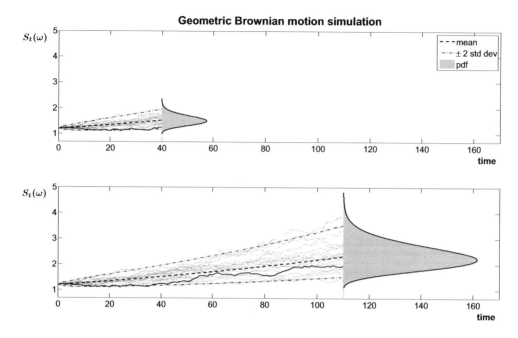

Figure 5.10: Evolution of a geometric Brownian motion

where $k = 1, \cdots, m$, $t_0 = 0$, $t_k - t_{k-1} = \Delta t \quad \forall k$, and $W_{t_k} = \sqrt{t_k} Z_{t_k}$ is the standard Brownian motion with $Z_{t_k} \sim N(0,1)$. Similar to Ex. 154, solve the following points.

1. Simulate a geometric Brownian motion with $n = 10000$, $m = 200$, $\mu = 0.2$, $\sigma = 0.2$, $\Delta t = 0.05$, and $S_0 = 100$.

2. Then, save the results in a .txt file.

3. Plot the trajectories of S_t with $t \in [0, 10]$.

4. Finally, plot the pdf of the geometric Brownian motion for fixed times $t_k = k\Delta t$ with $k = 10, 30, ..., 170, 190$, to show the evolution of the random variable S_{t_k}. Then, save these figures as .jpg files.

Hint: use the Function F_Empirical_pdf of Ex. 99.

Sol.: See Script S_GeoBM.

198

Chapter 6

Pricing of derivatives with an underlying security

In this chapter, using the probabilistic tools introduced earlier, we present three methodologies to price derivatives: the binomial model based on a discrete-time framework, the (continuous-time) Black-Scholes model, and the Monte Carlo Option model. More precisely, we mainly focus on pricing European Call and Put options. A European Call (Put) option is a contract, that gives to the buyer the right, but not the obligation, to purchase (sell) an underlying asset S at a specific future time T, i.e., the maturity of the option, for a fixed price K, named the strike price.

This chapter is structured as follows. In Section 6.1 we introduce the binomial model, describing its assumptions and the uniperiodal scheme of no-arbitrage for a replicating portfolio of stocks and bonds (Section 6.1.1). Then, in Section 6.1.2 we present the binomial model calibration of Cox-Ross-Rubinstein (CRR). Section 6.1.3 provides the CRR model in the multi-period case and proposes its practical application. In Section 6.2, we present the continuous time model for pricing derivatives introduced by Black and Scholes (1973); Merton (1973). For this, we first discuss the assumptions of the model, then in Section 6.2.2 we describe the dynamic of a call option. Section 6.2.3 provides the general equation of the Black-Scholes model for derivative pricing. In Section 6.2.4 we briefly discuss the concept of implied volatility. In Section 6.2.5 we show how to obtain the famous Black-Scholes formulas through integrals. In Section 6.3, exploiting the Feynman-Kaĉ formula, we describe the pricing of European call and put options through the Monte Carlo simulation, which is also used in Section 6.3.1 to evaluate path-dependent derivatives.

6.1 Binomial model

The binomial model for option pricing was proposed by Cox et al (1979). It can be considered the simplest model of no-arbitrage for the evaluation of derivatives. Such a model is defined on discrete time and considers only one source of uncertainty, namely the underlying price, that is modeled by a binomial process. In words, this means that, at the end of each time step Δt, the price S_t of the underlying can assume only two states, an upstate and a downstate, as shown in Fig. 6.1.

For simplicity, let us consider a single period time horizon $[0, T]$, and let S_0 be the price of a generic equity at time $t = 0$. Here we tackle with more details the evaluation of a European call that, at expiry T, has payoff $C_T = \max\{S_T - K, 0\}$. However, a similar procedure can be followed for pricing a European Put option that, at expiry T, has payoff $P_T = \max\{K - S_T, 0\}$.

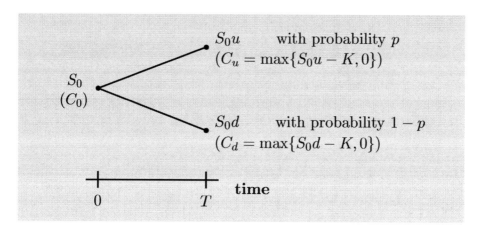

Figure 6.1: Scheme of an one-period binomial tree.

Furthermore, let us assume that the typical perfect-market hypotheses hold, namely

1. non-frictional market (e.g., no tax, no transaction costs);

2. operators are price takers and profit maximizers;

3. investors are risk averse;

4. no arbitrage opportunities.

Two additional key hypotheses are:

5. the stock does not pay dividends;

6. flat and deterministic term structure of interest rates $r(t, s) = r$.

This last hypothesis implies the absence of risk due to the interest rate fluctuations.

In the next section, in order to find the Call value C_0, we use the no-arbitrage principle by starting from a portfolio composed by stocks and bonds.

6.1.1 A replicating portfolio of stocks and bonds

In this section, we describe the replicating portfolio strategy in the case of one-period scheme as in Fig. 6.1. Let us denote the number of bonds and stocks in a portfolio by α_B and α_S, respectively, and the unit value of a Zero Coupon Bond (ZCB) at time $t = 0$ by $B_0 = 1$. By assumption 6 we have that for $T > 0$ $B_T = B_0 e^{rT} = e^{rT}$. On the other hand, we indicate the value of a stock at time $t = 0$ by S_0, while at time $T > 0$ by S_T. As discussed above, the model of Cox et al (1979) assumes that the time evolution of the asset price S_t is represented by a binomial process, and the possible future states of S_t are two as shown in Fig. 6.1.

Thus, the value of a portfolio Π composed by α_S stocks and α_B bonds at time $t = 0$ is

$$\Pi_0 = \alpha_S S_0 + \alpha_B B_0 = \alpha_S S_0 + \alpha_B$$

while, at time T, we have the following possible payoffs

$$\begin{cases} \Pi_u = \alpha_S u S_0 + \alpha_B e^{rT} & \text{with probability} \quad p \\ \\ \Pi_d = \alpha_S d S_0 + \alpha_B e^{rT} & \text{with probability} \quad 1 - p \end{cases}$$

where u is the multiplicative coefficient related to the upward state, while d is the one related to the downward state. The structure of the model works if $d < u$. In order to exploit the *law of the one price* (i.e., in a condition of no arbitrage, two contracts with the same payoffs for each due date must have the same price), we choose α_B and α_S to replicate the payoffs of the Call option C_u and C_d at time T. Then, we obtain the following system of linear equations:

$$\begin{cases} \alpha_S u S_0 + \alpha_B e^{rT} = C_u \\ \\ \alpha_S d S_0 + \alpha_B e^{rT} = C_d \,. \end{cases} \tag{6.1}$$

The solution to (6.1) is

$$
\alpha_S^\star = \frac{\begin{vmatrix} C_u & e^{rT} \\ C_d & e^{rT} \end{vmatrix}}{\begin{vmatrix} uS_0 & e^{rT} \\ dS_0 & e^{rT} \end{vmatrix}} = \frac{e^{rT}\left(C_u - C_d\right)}{e^{rT}S_0\left(u - d\right)} = \frac{\left(C_u - C_d\right)}{S_0\left(u - d\right)} \tag{6.2}
$$

$$
\alpha_B^\star = \frac{\begin{vmatrix} uS_0 & C_u \\ dS_0 & C_d \end{vmatrix}}{\begin{vmatrix} uS_0 & e^{rT} \\ dS_0 & e^{rT} \end{vmatrix}} = \frac{S_0\left(uC_d - dC_u\right)}{e^{rT}S_0\left(u - d\right)} = \frac{\left(uC_d - dC_u\right)}{e^{rT}\left(u - d\right)} . \tag{6.3}
$$

According to the *law of the one price* and given the conditions (6.1), we have that at time $t = 0$ the price of the Call C_0 and of the portfolio Π_0 with shares (6.2) and (6.3) must be equal

$$
C_0 = \Pi_0 = \alpha_S^\star S_0 + \alpha_B^\star . \tag{6.4}
$$

Therefore, substituting (6.2) and (6.3) in (6.4), we obtain

$$
\begin{aligned}
C_0 &= \frac{C_u - C_d}{S_0(u - d)}S_0 + e^{-rT}\frac{uC_d - dC_u}{u - d} \\
&= e^{-rT}\frac{e^{rT}C_u - e^{rT}C_d + uC_d - dC_u}{u - d} \\
&= e^{-rT}\left[\frac{e^{rT} - d}{u - d}C_u + \frac{u - e^{rT}}{u - d}C_d\right] .
\end{aligned} \tag{6.5}
$$

If we denote $q = \dfrac{e^{rT} - d}{u - d}$ and $1 - q = \dfrac{u - e^{rT}}{u - d}$, then we can rewrite Eq. (6.5) as

$$
C_0 = e^{-rT}\left[q\,C_u + (1 - q)\,C_d\right] . \tag{6.6}
$$

Note that, if $d < e^{rT} < u$ holds, then q and $1 - q$ are always positive and lower than one. Thus, q and $1 - q$ can be interpreted as probabilities related to the upstate and to the downstate, respectively, and, therefore, the terms of Expression (6.6) in brackets can be regarded as the expected value of the future payoffs of the call option.

The conditions $d < e^{rt} < u$ must be verified, otherwise arbitrage possibilities can occur. Indeed, if $e^{rt} < d < u$, one could short sell α unit ZCBs and buy α stocks of unit value, determining arbitrage opportunities by the following payoff table:

position	$t = 0$	$t = 1$
short sell α ZCBs	$\alpha \cdot 1$	$-\alpha e^{rt}$
buy α stocks	$-\alpha \cdot 1$	αu
		αd
	0	$\alpha(u - e^{rt}) > 0$
		$\alpha(d - e^{rt}) > 0$

Similarly, if $d < u < e^{rt}$, then one could short sell α stocks of unit value and buy α unit ZCBs, thus generating arbitrage opportunities.

Remark 163 (Feynman-Kaĉ formula) *Expression (6.6) for the price of a call option could be considered as the discounted expected value of the future payoffs of the call. More precisely, we can write that*

$$C_0 = e^{-rT} \left(q\, C_u + (1 - q)C_d \right) = e^{-rT} E_0^{\mathbb{Q}} \left[C_T \right]. \tag{6.7}$$

In the general case, this result is the Feynman-Kaĉ formula, and $\mathbb{Q}$ indicates that the expectation is done under the risk-neutral probability measure, whose intuitive meaning will be mentioned in the next remark. Thus, q and $1 - q$ can be interpreted as the risk-neutral probabilities associated to the future payoffs of the call, C_u and C_d, respectively. Note that the risk-neutral probabilities are different from p and $1 - p$ (see Fig. 6.1), which are the real (or natural) probabilities. Generally, the real probability measure is indicated by $\mathbb{P}$.

Remark 164 (Few words about $\mathbb{Q}$) *Here we better specify the meaning of the risk-neutral probability measure $\mathbb{Q}$. In the framework of the expected utility theory, the operators are generally profit maximizers and risk averse, and their satisfaction w.r.t. the wealth w is represented by a utility function $u(w)$, where $u'(w) > 0$ (i.e., profit maximizers) and $u''(w) < 0$ (i.e., risk averse). Then, the evaluation of a random payoff C_T at a future time T, realized by a generic operator, requires discounting the certainty equivalent of C_T, denoted by $\widetilde{C}_T$, by means of the risk-free rate. More specifically, we have that the certainty equivalent $\widetilde{C}_T$ is such that $u\left(\widetilde{C}_T\right) = E_0\left[u\left(C_T\right)\right]$. This implies that*

$$\widetilde{C}_T = u^{-1} \left(E_0 \left[u \left(C_T \right) \right] \right) = u^{-1} \left(p u(C_u) + (1 - p)u(C_d) \right),$$

and, thus, in the framework of the expected utility theory, we have

$$C_0 = e^{-rT} \widetilde{C}_T = e^{-rT} u^{-1} \left(p u(C_u) + (1 - p)u(C_d) \right). \tag{6.8}$$

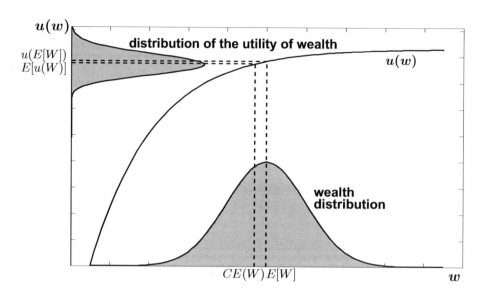

$u(w)$

$u(E[W])$
$E[u(W)]$

distribution of the utility of wealth

$u(w)$

wealth distribution

$CE(W)\,E[W]$

w

Figure 6.2: Example of the pdf of a random wealth (horizontally) and the pdf of the same random wealth deformed by a utility function of a risk-averse operator (vertically)

According to the hypotheses of risk averse investors, u is a concave function, namely $\alpha\,u(w_x) + \beta\,u(w_y) \leq u(\alpha w_x + \beta w_y)$. As graphically shown in Fig. 6.2, this implies that the following relation holds

$$u^{-1}\left(\,pu(C_u) + (1-p)u(C_d)\,\right) \leq p\,C_u + (1-p)C_d\,. \qquad (6.9)$$

Note that in the case of risk lover operators Relation (6.9) is inverted, as represented in Fig. 6.3. The equality is obtained when the utility function u is linear, namely when the operators are indifferent w.r.t risk, i.e., when they are risk-neutral (see Fig. 6.4). From this observation, the risk-neutral *definition is originated. Thus, if the interpretation of q and $1-q$ as probabilities is accepted, then $C_0 = e^{-rT}E_0^{\mathbb{Q}}\,[\,C_T\,]$ takes the role of the discounted certainty equivalent defined in Eq. (6.8), when the operators are risk-neutral.*
Note that the probabilities of Eq. (6.8), i.e., p and $1-p$, represent the real (or natural) probabilities, which, following the De Finetti's definition (see Section 2.1), are subjective, as well as the utility function u. Conversely, Eq. (6.7) is independent from the preferences of the operator, and therefore the probabilities q and $1-q$ can be interpreted as the risk-neutral (or risk-adjusted) probabilities.

Although the single period time horizon with two state of nature (see Fig. 6.1) is a relevant and instructive case, to better represent the uncertainty of the

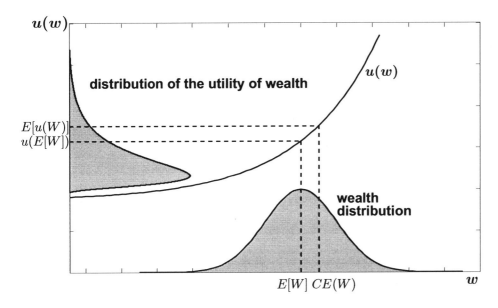

Figure 6.3: Example of the pdf of a random wealth (horizontally) and the pdf of the same random wealth deformed by a utility function of a risk-lover operator (vertically).

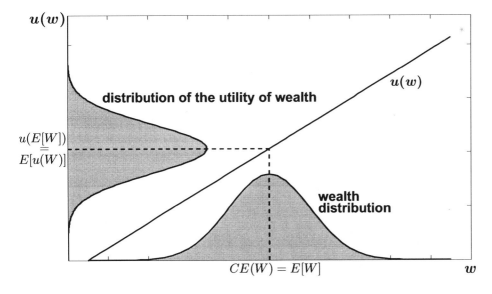

Figure 6.4: Example of the pdf of a random wealth (horizontally) and the pdf of the same random wealth deformed by a utility function of a risk-neutral operator (vertically).

underlying price, a multi-period binomial tree should be considered, as represented in Fig. 6.5. Furthermore, the choice of using a recombining binomial

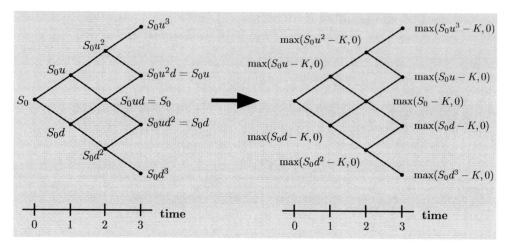

Figure 6.5: Example of a recombining binomial tree

tree is due to the computational and efficiency convenience. To obtain this *recombining* binomial tree, it is necessary to impose $u = \dfrac{1}{d}$, and, therefore, we have $S_0 u d = S_0$.

In the following section, we describe how to calibrate the binomial model. In other word, we show how to set the parameters of the CRR model, namely the risk-neutral probability q, and the multiplicative coefficients u and d related to the upward and downward state, respectively.

6.1.2   Calibration of the binomial model

Model calibration is a procedure that consists of fine-tuning the values of the input parameters in order to match the real-world conditions as best as possible. In other words, the calibration process involves the adjustment of the input parameters of the model, so that the outputs of the model can accurately reflect some specific benchmarks. For instance, the calibration could ask for minimizing the difference between the theoretical results obtained by the model and the historical data given by the market, using several statistical techniques (e.g., by means of the maximum likelihood estimation). On the other hand, the stochastic process for the price of a stock generated by the binomial model (namely the binomial process) should maintain some properties of its continuous version, i.e., the geometric Brownian motion (used in the Black-Scholes model, see Section

6.2). The Cox-Ross-Rubinstein (CRR) calibration is based on this approach (see Cox et al, 1979).

As shown in Section 5.2.3, the geometric Brownian motion can be described by the following stochastic differential equation $dS_t = \mu S_t dt + \sigma S_t dW_t$. It is possible to prove that, under the risk-neutral (or risk-adjusted) probability measure $\mathbb{Q}$, the drift of the geometric Brownian motion has to be equal to the risk-free interest rate r, thus becoming $dS_t = r S_t dt + \sigma S_t dW_t$. Furthermore, according to (5.25), given a discrete time increment Δt, the log-return between S_t and $S_{t+\Delta t}$ is normally distributed as follows

$$\log\left(\frac{S_{t+\Delta t}}{S_t}\right) \sim N\left(\left(r - \frac{\sigma^2}{2}\right)\Delta t, \sigma^2 \Delta t\right), \tag{6.10}$$

namely $\frac{S_{t+\Delta t}}{S_t} \sim \ln N\left(\left(r - \frac{\sigma^2}{2}\right)\Delta t, \sigma^2 \Delta t\right)$. This implies that from (2.27) and (2.28) we have

$$E_t^{\mathbb{Q}}\left[\frac{S_{t+\Delta t}}{S_t}\right] = e^{r\Delta t}$$

$$Var_t^{\mathbb{Q}}\left[\frac{S_{t+\Delta t}}{S_t}\right] = \left(e^{\sigma^2 \Delta t} - 1\right)e^{2r\Delta t}.$$

However, since S_t is known in t, we can write

$$E_t^{\mathbb{Q}}\left[S_{t+\Delta t}\right] = S_t e^{r\Delta t} \tag{6.11}$$

$$Var_t^{\mathbb{Q}}\left[S_{t+\Delta t}\right] = S_t^2\left(e^{\sigma^2 \Delta t} - 1\right)e^{2r\Delta t}. \tag{6.12}$$

To calibrate the model, we impose that the first two moments of the binomial process at time t are equal to those of the Geometric Brownian process. For the binomial process we have

$$E_t^{\mathbb{Q}}\left[S_{t+\Delta t}\right] = q\,uS_t + (1-q)dS_t \tag{6.13}$$

$$Var_t^{\mathbb{Q}}\left[S_{t+\Delta t}\right] = E_t^{\mathbb{Q}}\left[S_{t+\Delta t}^2\right] - E_t^{\mathbb{Q}}\left[S_{t+\Delta t}\right]^2$$

$$= q\,u^2 S_t^2 + (1-q)d^2 S_t^2 - S_t^2 e^{2r\Delta t}. \tag{6.14}$$

Thus making (6.11) and (6.13) equal, we obtain that

$$S_t e^{r\Delta t} = q\,uS_t + (1-q)dS_t$$

$$\Rightarrow \quad q = \frac{e^{r\Delta t} - d}{u - d}, \tag{6.15}$$

while imposing the equality between (6.12) and (6.14), we have

$$S_t^2 e^{2r\Delta t}\left(e^{\sigma^2 \Delta t} - 1\right) = S_t^2\left[qu^2 + (1-q)d^2 - e^{2r\Delta t}\right]$$

$$\Rightarrow \qquad e^{2r\Delta t + \sigma^2 \Delta t} = qu^2 + (1-q)d^2 . \tag{6.16}$$

Now, substituting (6.15) in (6.16), we can write

$$
\begin{aligned}
e^{2r\Delta t + \sigma^2 \Delta t} &= \frac{e^{r\Delta t} - d}{u - d}u^2 + \frac{u - e^{r\Delta t}}{u - d}d^2 \\
&= \frac{1}{u - d}\left[e^{r\Delta t}u^2 - du^2 + ud^2 - e^{r\Delta t}d^2\right] \\
&= \frac{(u - d)(u + d)}{u - d}e^{r\Delta t} - \frac{du(u - d)}{u - d} \\
&= (u + d)e^{r\Delta t} - 1 , \tag{6.17}
\end{aligned}
$$

where, in the last step, since the lattice is recombining, we have that $ud = 1$. Again, since $d = \dfrac{1}{u}$,

$$e^{2r\Delta t + \sigma^2 \Delta t} = (u + \frac{1}{u})e^{r\Delta t} - 1$$

$$\Rightarrow \qquad ue^{2r\Delta t + \sigma^2 \Delta t} = (u^2 + 1)e^{r\Delta t} - u$$

$$\Rightarrow \qquad u^2 e^{r\Delta t} - u\left(e^{2r\Delta t + \sigma^2 \Delta t} + 1\right) + e^{r\Delta t} = 0$$

$$\Rightarrow \quad u_{1,2} = \frac{\left(e^{2r\Delta t + \sigma^2 \Delta t} + 1\right) \pm \sqrt{\left(e^{2r\Delta t + \sigma^2 \Delta t} + 1\right)^2 - 4e^{2r\Delta t}}}{2e^{r\Delta t}} . \tag{6.18}$$

Considering a small Δt and neglecting the terms of order higher than one w.r.t. Δt, the exponential function can be approximated by its Maclaurin[1] expansion as $e^x \simeq 1 + x$. Thus, we can approximate Expression (6.18) as follows

---

[1]The Maclaurin series is a special case of the Taylor series where $x_0 = 0$.

$$u_{1,2} \simeq \frac{1 + 2r\Delta t + \sigma^2\Delta t + 1 \pm \sqrt{(1 + 2r\Delta t + \sigma^2\Delta t + 1)^2 - 4(1 + 2r\Delta t)}}{2e^{r\Delta t}}$$

$$\simeq \frac{2 + 2r\Delta t + \sigma^2\Delta t \pm \sqrt{4 + 8r\Delta t + 4\sigma^2\Delta t - 4 - 8r\Delta t}}{2}e^{-r\Delta t}$$

$$\simeq \frac{2\left(1 + r\Delta t + \frac{\sigma^2}{2}\Delta t \pm \sigma\sqrt{\Delta t}\right)(1 - r\Delta t)}{2}$$

$$\simeq 1 + r\Delta t + \frac{\sigma^2}{2}\Delta t \pm \sigma\sqrt{\Delta t} - r\Delta t \underbrace{-r^2\Delta t^2 - \frac{\sigma^2}{2}r\Delta t^2 \mp \sigma r\Delta t^{\frac{3}{2}}}_{\approx 0}$$

$$\simeq 1 \pm \sigma\sqrt{\Delta t} + \frac{\sigma^2}{2}\Delta t . \tag{6.19}$$

Therefore, $u_{1,2} \simeq 1 \pm \sigma\sqrt{\Delta t} + \frac{\sigma^2}{2}\Delta t$ has the form of the Maclaurin expansion of $e^{\pm\sigma\sqrt{\Delta t}}$, but since u has to be greater than 1, $u = u_1$. Then, we can set

$$u = e^{\sigma\sqrt{\Delta t}}, \quad d = e^{-\sigma\sqrt{\Delta t}}, \quad q = \frac{e^{r\Delta t} - d}{u - d},$$

which represents the CRR parametrization.

6.1.3 Multi-period case

In this section, we consider a European call option on an underlying stock S having maturity T and strike price K, where the uncertainty of the underlying is described by a multi-period recombining binomial tree as in Fig. 6.5. Furthermore, as mentioned at the beginning of this chapter, the payoff at expiry is $C(T) = \max(S(T) - K, 0)$.

Let us denote the value of the call option at the generic node (k, j) of the recombining binomial tree by C_{kj}, where j indicates the discretization of time, namely $t = j\Delta t$, with $j = 0, 1, \ldots, N$, while k indicates the state at time j, with $k = 0, 1, \ldots, j$. Since N represents the number of time steps, $\Delta t = \frac{T}{N}$.

Using this discretization of time, the price of a stock S_{kj} in the node (k, j) is $S_{kj} = S_0 u^k d^{j-k}$ with $j = 0, 1, \ldots, N$ and $k = 0, 1, \ldots, j$, as shown in Fig. 6.6. In particular, at maturity T, we have

$$S(T) = \{S_{k,N}\}_{k=0,1,\ldots,j=N} = \left\{S_0 u^k d^{N-k}\right\}_{k=0,1,\ldots,N}.$$

As a consequence, the payoff of the call at maturity T is

$$C(T) = \{C_{k,N}\}_{k=0,1,\ldots,N} = \left\{\max\{S_0 u^k d^{N-k} - K, 0\}\right\}_{k=0,1,\ldots,N}. \tag{6.20}$$

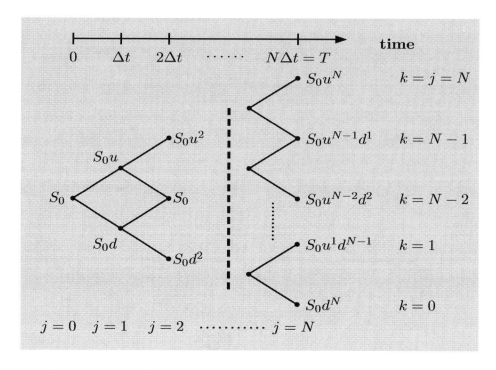

Figure 6.6: Realizations of the equity price in the binomial lattice.

As shown in Fig. 6.7, applying to each node (k, j) (with $j = 0, \ldots, N - 1$ and $k = 0, \ldots, j$) the Feynman-Kaĉ Formula (6.7) for each time step Δt (i.e., $C_t = e^{-r\Delta t} E_t^{\mathbb{Q}}[C_{t+\Delta t}]$), we have

$$C_{k,j} = e^{-r\Delta t} \left(q\, C_{k+1,j+1} + (1 - q) C_{k,j+1} \right) . \tag{6.21}$$

Hence, to find the value of the call in $t = 0$, namely $C_0 = C_{0,0}$, a backward procedure can be used until we reach the node $(j = 0, k = 0)$. Such a procedure is also an operational way to price the option. For more details, see Exercise 165.

Exercise 165 (Pricing of a European Option) *Let us consider a European call and put option on an underlying stock with a current price $S_0 = 100$, with maturity $T = 1$ year and a strike price $K = 100$. Furthermore, we know that the annual risk-free interest rate is $i = 0.05$, and that the volatility of the underlying stock is $\sigma = 0.20$. Consider a binomial model with $N = 50$ time steps, and compute*

$$\Delta t = T/N , \qquad r = \ln(1 + i) .$$

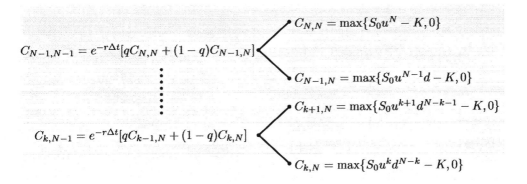

Figure 6.7: Evaluation of a call using a backward procedure on a recombining binomial tree.

According to the CRR parametrization (see Section 6.1.2), the multiplicative upstate and downstate coefficients are respectively

$$u = e^{\sigma\sqrt{\Delta t}}, \qquad d = 1/u,$$

and the risk-neutral probabilities are

$$q = \frac{e^{r\Delta t} - d}{u - d}, \qquad 1 - q,$$

related to the upstate and to the downstate, respectively. Using a backward procedure on a recombining binomial tree, evaluate the price of a European call and a European put. For this aim, write a Function (F_Binomial) with strike price, current price, volatility of the underlying, maturity, risk free rate and the number of steps as inputs, and with the prices of the call and the put as outputs. Furthermore, compare and check the price of the put obtained from the put-call-parity relationship:

$$P_t = C_t - S_t + K\,e^{-r(T-t)}.$$

Thus using the input data described above, write a Script named S_Binomial, where you recall the Function F_Binomial.

Sol.: See Function F_Binomial and Script S_Binomial.

However, it is also possible to find an explicit expression of $C(t = 0) = C_{0,0}$. Indeed, looking at Fig. 6.7 and using (6.21), one can note that $C_{k,N-1}$, namely the values of the call at time $j = N - 1$, can be written as a function of $C_{k+1,N}$ and $C_{k,N}$, whose explicit expression is given by (6.20). Thus proceeding backward (going back in time from the maturity T towards $t = 0$) and appropriately substituting the values of $C_{k,j}$ to the nodes with $j - 1$, one can achieve the following result

$$C(0) = e^{-rN\Delta t} \left(\sum_{k=0}^{N} \frac{N!}{k!(N-k)!} q^k (1-q)^{N-k} \max\{ u^k d^{N-k} S_0 - K, 0 \} \right),$$
(6.22)

where $p(k, N; q) = \dfrac{N!}{k!(N-k)!} q^k (1-q)^{N-k}$ is the pdf of a binomial random variable and represents the probability that, after N steps, k upward states occur, given that q is the probability of one upward.

Following the same steps and considerations used to find the call option price (6.22), given the boundary conditions $P_T = \max\{K - S_T, 0\}$, one can find the following explicit form for a European put option:

$$P(0) = e^{-rN\Delta t} \left(\sum_{k=0}^{N} \frac{N!}{k!(N-k)!} q^k (1-q)^{N-k} \max\{ K - u^k d^{N-k} S_0, 0 \} \right).$$
(6.23)

Remark 166 *Note that if we denote by h the value of the index k such that the option value is greater than zero, then we obtain that*

$$C(T) = \{C_{k,N}\}_{\forall k} \quad = \{\max\{u^k d^{N-k} S - K, 0\}\}_{\forall k} \quad (6.24)$$
$$= \begin{cases} u^k d^{N-k} S - K & \text{if } k \geq h \\ 0 & \text{if } k < h \end{cases}$$

Then, h can be found as follows:

$$u^h d^{N-h} S_0 - K = 0 \quad \Rightarrow$$
$$\Rightarrow \quad \left(\frac{u}{d}\right)^h d^N = \frac{K}{S_0} \quad \Rightarrow$$
$$\Rightarrow \quad h \ln\left(\frac{u}{d}\right) = \ln \frac{K}{S_0} - \ln d^N \quad \Rightarrow$$
$$\Rightarrow \quad h = \left\lfloor \frac{\ln\left(\dfrac{K}{S_0 d^N}\right)}{\ln\left(\dfrac{u}{d}\right)} \right\rfloor \quad (6.25)$$

where the symbol $\lfloor x \rfloor$ represents the flooring function that maps x to the nearest integer towards minus infinity. Thus, we can write that

$$C(t = 0) = C_{0,0} = e^{-rN\Delta t} \sum_{k=h}^{N} \frac{n!}{k!(N-k)!} q^k (1-q)^{N-k} \left(u^k d^{N-k} S - K \right) \quad (6.26)$$

$$P(t=0) = P_{0,0} = e^{-rN\Delta t} \sum_{k=0}^{h} \frac{n!}{k!(N-k)!} q^k (1-q)^{N-k} \left(K - u^k d^{N-k} S \right) \quad (6.27)$$

Using (6.25), (6.26) and (6.27) to price call and put options, we can obtain a slightly faster procedure than that achieved by means of Expressions (6.22) and (6.23).

Remark 167 (CRR model vs. Black-Scholes model) *It can be proved that, considering a time step $\Delta t \to 0$ (or equivalently, $N \to \infty$) and using the CRR parametrization, the binomial process for modeling S_t tends to the geometric Brownian motion with a drift equal to r and a diffusion coefficient equal to σ. Furthermore, Expressions (6.22) and (6.23) tend to the Black-Scholes formulas (see Section 6.2).*

Exercise 168 (Pricing of a European Option 2) *Given the same market data of Exercise 165, evaluate the price of a European call and a European put, using the binomial formulas (6.26) and (6.27). For this aim, write a Function (F_Binomial2) with strike price, current price, volatility of the underlying, maturity, risk free rate and the number of steps as inputs, and with the prices of the call and the put as outputs. Furthermore, compare and check the price of the put obtained from the put-call-parity relationship:*

$$P_t = C_t - S_t + K\, e^{-r(T-t)}\,.$$

Thus considering the input data reported in Exercise 165, write a Script (named S_Binomial2) in which you recall the Function F_Binomial2. Clearly, the results must be the same.

Furthermore, appropriately using the Function F_Binomial of Exercise 165, plot the price of the call option as a function of the number N of time steps considered, e.g., $N = 1, \ldots, 100$.

Sol.: *See Function F_Binomial2 and Script S_Binomial2.*

6.2 Black-Scholes model

In this section, we present the continuous time model for pricing derivatives, introduced by Black and Scholes (1973); Merton (1973). For this, we first discuss the assumptions of the model, then in Section 6.2.2 we describe the dynamic of a call option. Section 6.2.3 provides the general equation of the Black-Scholes model for derivative pricing. In Section 6.2.4 we briefly discuss the concept of implied volatility. In Section 6.2.5 we present how to obtain the famous Black-Scholes formula through integrals.

6.2.1 Assumptions of the model

We discuss here the assumptions of the Black-Scholes model to evaluate the no-arbitrage price of a derivative contract.

Assume that $t = 0$ is the inception, $t = T$ is the maturity of the contract, and $t \in [0, T]$ is a generic intermediate time. Similar to the assumptions of the binomial model, described in Section 6.1, the hypothesis of the Black-Scholes model are the following:

1. perfect market:

    - no transaction costs and no taxes;
    - the equities are infinitely divisible;
    - short-selling is allowed;
    - the operators are price-takers and profit maximizers;

2. no arbitrage opportunities;

3. the market provides Zero Coupon Bond (ZCB) with any maturity, and the term structure of interest rates is flat and deterministic, with an instantaneous intensity of interest equal to r;

4. the market is always open;

5. the underlying price S_t, with $t \in [0, T]$, follows a geometric Brownian motion, namely

$$dS_t = \mu S_t dt + \sigma S_t dW_t, \qquad (6.28)$$

    where μ and σ are scalars, and $S_0 = s_0$ (see Section 5.2.3);

6. the underlying S does not pay any dividend.

In the following sections we provide a detailed description for some of the above hypothesis.

Bond market: flat and deterministic term structure

Let $v(t, s)$ denote the price of a unit ZCB. In terms of instantaneous intensity of interest $r(t, s)$, we have

$$v(t, s) = e^{-\int_t^s r(t,u)du} \, .$$

The assumption of flat term structure of interest rates implies that $r(t, s) = r$ with $t \leq s$, while given the assumption of deterministic term structure of interest rates we have that

$$v(\tau, s) = \frac{v(t, s)}{v(t, \tau)} \qquad \forall \, \tau \in [t, s].$$

These two assumptions together imply that

$$v(\tau, s) = e^{-r(s-\tau)} \qquad \forall \, \tau \in [t, s],$$

where the scalar r is the yield to maturity, that coincides with the instantaneous intensity of interest in the case of flat and deterministic term structure.

Remark 169 (Money market account) *We assume the existence of a money market account, which is a bond fund characterized by investments of very short term. If $B(0)$ is a capital invested in $t = 0$ in the money market account, then, at a generic time t, the value of the bond investment $B(t)$ will be*

$$B(t) = B(0)e^{rt}, \qquad (6.29)$$

where $t \geq 0$. Thus differentiating Expression (6.29), we obtain

$$
\begin{aligned}
dB(t) &= B(0)e^{rt}r\,dt \\
dB(t) &= rB(t)dt,
\end{aligned}
\qquad (6.30)
$$

where $r = \frac{dB(t)}{B(t)dt}$ represents the instantaneous intensity of interest, and the deterministic differential equation (6.30) describes the dynamics of an investment $B(t)$ in the money market account.

Stocks Market

As mentioned in Section 6.2.1, the Black-Scholes model assumes that the price S_t of the underlying asset follows a geometric Brownian motion. Similarly, we can also say that the infinitesimal return dS_t/S_t evolves according to the following arithmetic Brownian motion (see Remark 155):

$$\frac{dS_t}{S_t} = \mu dt + \sigma dW_t, \qquad (6.31)$$

where μdt is the deterministic part of the process with $\mu > r$. The second term σdW_t is the random part of the process, where σ, by assumption, is constant, although in the real markets we observe the so-called *volatility smile* effect (see, e.g., Hull and Basu, 2016, and references therein). This latter phenomenon shows that σ is not constant in practice, but it depends on the maturity T and on the strike price K of the option. Furthermore, $dW_t \sim N(0, dt)$, while on the real markets the assets random returns generally show a probability distribution with fatter tails than those of a Gaussian distribution (Cont, 2001). Typically, a leptokurtic distribution (e.g., Student-t) can guarantee a better representation of the equity return.

In order to better understand the financial meaning of μ and σ, let us now consider the expectation of Eq. (6.31), namely

$$E_t \left[\frac{dS_t}{S_t} \right] = \mu dt + \sigma E_t \left[dW_t \right]$$

$$= \mu dt$$

$$\Rightarrow \mu = \frac{E_t \left[\frac{dS_t}{S_t} \right]}{dt}.$$

Hence, μ can be viewed as the instantaneous intensity of return of the stock price. Furthermore, one can also compute the variance of (6.31) as follows

$$Var_t \left[\frac{dS_t}{S_t} \right] = \sigma^2 Var_t \left[dW_t \right] = \sigma^2 dt$$

$$\Rightarrow \sigma^2 = \frac{Var_t \left[\frac{dS_t}{S_t} \right]}{dt}.$$

Thus, σ represents the standard deviation of the instantaneous return of the stock price.

Remark 170 (Sharpe ratio) *From the Expected Utility theory we recall that if $\mu > r$, then $\mu - r$ represents the risk premium. Furthermore, considering σ as a risk measure, one can compute the following ratio*

$$\pi_S = \frac{\mu - r}{\sigma},$$

which represents the premium earned per unit of risk, or, in other words, the market price of risk. Note that π_S is commonly defined as the continuous Sharpe ratio of a stock.

Remark 171 *Recall that if the price of a stock follows a geometric Brownian motion, then $S_T \sim \ln N\left(\ln s_0 + (\mu - \frac{\sigma^2}{2})T, \sigma^2 T\right)$. For this reason, one can define the probability $P\left(S_T \in (a,b)\right)$, which, in our framework, is a subjective probability. Then, since the type of distribution is fixed by hypothesis (i.e., log-normal), the only subjective elements are represented by the parameters μ and σ. Generally, the estimation error on μ is greater than that on σ. However, as we shall see in Section 6.2.3, the drift μ of the geometric Brownian motion, that represents the underlying price of an option, does not appear in the BlackScholes equation (6.41), namely the option price does not depend on the expected (instantaneous) return of the underlying (Black and Scholes, 1973; Scholes, 1998).*

Remark 172 *In the following section, assuming that there are not dividends for the underlying between t and T, we show the Black-Scholes model applied to the European call and put option pricing. However, similar argumentations can be used to price different derivatives with various payoffs, as shown in Fig. 6.8.*

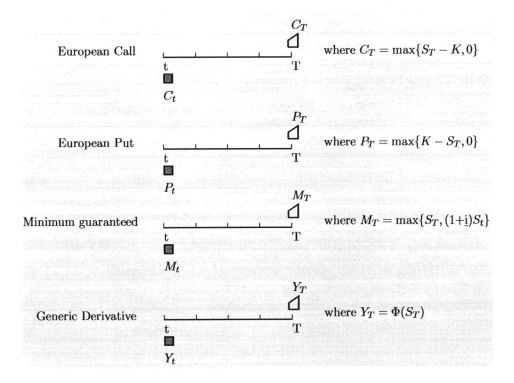

Figure 6.8: Example of some common derivatives.

6.2.2 Pricing of a European call

As for the binomial model described in Section 6.1, the Black-Scholes approach is based on the principle of no arbitrage opportunities. Specifically, in a dynamic context (introduced in the next section) the Black-Scholes model uses a hedging argumentation that instantaneously satisfies the constraint of absence of arbitrage. This scheme leads us to determine the famous Black-Scholes partial differential equation (6.41) for pricing derivatives.

The dynamics of the option price

Relying on the assumptions described in Section 6.2.1, we now turn to examine the evolution of the no-arbitrage price of an option.

Let us denote the price of the European call at time t as $C_t = C(S_t, t)$, with $dS_t = \underbrace{\mu S_t}_{\mu_t} dt + \underbrace{\sigma S_t}_{\sigma_t} dW_t$. Therefore, it is possible to apply Ito's Lemma (5.10) to evaluate infinitesimal increments of the call price as follows:

$$dC_t = \left[\frac{\partial C}{\partial t} + \frac{\partial C}{\partial S_t} \mu_t + \frac{1}{2} \frac{\partial^2 C}{\partial S_t^2} \sigma_t^2 \right] dt + \frac{\partial C}{\partial S_t} \sigma_t dW_t$$

$$dC_t = \left[\frac{\partial C}{\partial t} + \frac{\partial C}{\partial S_t} \mu S_t + \frac{1}{2} \frac{\partial^2 C}{\partial S_t^2} \sigma^2 S_t^2 \right] dt + \frac{\partial C}{\partial S_t} \sigma S_t dW_t$$

$$dC_t = \tilde{a}_t dt + \tilde{b}_t dW_t, \qquad (6.32)$$

where $\tilde{a}_t = \frac{\partial C}{\partial t} + \frac{\partial C}{\partial S_t} \mu S_t + \frac{1}{2} \frac{\partial^2 C}{\partial S_t^2} \sigma^2 S_t^2$ and $\tilde{b}_t = \frac{\partial C_t}{\partial S_t} \sigma S_t$. Let us now define the return of C generated between t and $t + dt$ as follows

$$\frac{dC_t}{C_t} = a_t dt + b_t dW_t, \qquad (6.33)$$

where $a_t = \frac{\tilde{a}_t}{C_t}$ and $b_t = \frac{\tilde{b}_t}{C_t}$. Considering the expected value and the variance of Eq. (6.33), we have

$$E_t \left[\frac{dC_t}{C_t} \right] = a_t dt \ \Rightarrow$$

$$\Rightarrow \ a_t = \frac{E_t \left[\frac{dC_t}{C_t} \right]}{dt},$$

218

where, therefore, a_t represents the expected value of the instantaneous intensity of return obtained by investing in the option. Furthermore, we have

$$Var_t \left[\frac{dC_t}{C_t} \right] = b_t^2 Var_t \left[dW_t \right] = b_t^2 dt$$

$$\Rightarrow \quad b_t^2 = \frac{Var_t \left[\frac{dC_t}{C_t} \right]}{dt},$$

consequently, b_t is the standard deviation of the instantaneous intensity of return. Thus, as in the case of the underlying price S (see Remark 170), one can define the *market* price of risk of the call C. Assuming that $a_t > r$, we can define the risk premium as $a_t - r$. Therefore, the risk market price of C is

$$\pi_c = \frac{a_t - r}{b_t},$$

where π_c represents the continuous Sharpe ratio of the call C.

A hedging strategy

Given a generic time t, one can determine the no-arbitrage price C_t of a call option (or, more in general, of a derivative Y_t, see Fig. 6.8) using a hedging strategy. More precisely, we can generate an instantaneously risk-free portfolio consisting of a call option C and its underlying S. Exploiting the no-arbitrage condition, the dynamic of this hedged portfolio must be equal to the dynamic of a bond investment in the money market account (see Remark 169).

We then consider a portfolio composed of a call option C and α units of its underlying S. At time t, the value of this portfolio is $\Pi_t = C_t + \alpha S_t$. The dynamics of Π_t can be described by the following stochastic differential equation

$$d\Pi_t = \Pi_{t+dt} - \Pi_t = (C_{t+dt} - C_t) + \alpha (S_{t+dt} - S_t)$$
$$\Rightarrow d\Pi_t = dC_t + \alpha dS_t. \tag{6.34}$$

Substituting (6.28) and (6.32) in Eq. (6.34), we have

$$d\Pi_t = \left(\widetilde{a}_t \, dt + \widetilde{b}_t \, dW_t \right) + \alpha \left(\mu S_t \, dt + \sigma S_t \, dW_t \right)$$
$$\Rightarrow d\Pi_t = \left(\widetilde{a}_t + \alpha \mu S_t \right) dt + \left(\widetilde{b}_t + \alpha \sigma S_t \right) dW_t. \tag{6.35}$$

Note that if $\alpha = \alpha^* = -\dfrac{\widetilde{b}_t}{\sigma S_t}$ the random term in (6.35) becomes equal to zero, and the portfolio is instantaneously risk-free. Then, the value of the instantaneously risk-free portfolio at time t is

$$\Pi_t^* = C_t + \alpha^* S_t,$$

and its dynamics is described by the following (now deterministic) differential equation

$$d\Pi_t^* = (\tilde{a}_t + \alpha^* \mu S_t)\, dt \ .$$

This means that Π_{t+dt}^* is perfectly predictable in t.

Remark 173 (Instantaneously risk-free portfolio) *Note that the portfolio Π_t is risk-free between t and $t + dt$ only for $\alpha = \alpha^*$. Furthermore, by the no-arbitrage condition, the portfolio $\Pi_t^* = C_t + \alpha^* S_t$, between t and $t + dt$, must generate the same interest obtained by an investment of a monetary amount Π_t^* in the bond market, or, more precisely, in the money market account (see Remark 169) with a fixed risk-free rate r. Therefore, we have*

$$
\begin{aligned}
d\Pi_t^* &= \Pi_t^* r\, dt \\
\Rightarrow (\tilde{a} + \alpha^* \mu S)\, dt &= \Pi_t^* r\, dt = (C_t + \alpha^* S_t)\, r\, dt \,,
\end{aligned}
\tag{6.36}
$$

where $\alpha^* = -\dfrac{\tilde{b}_t}{\sigma S_t}$. Thus, from (6.36) we obtain

$$
\begin{aligned}
\tilde{a}_t - \mu \frac{\tilde{b}_t}{\sigma} &= rC_t - r\frac{\tilde{b}_t}{\sigma} \\
\Rightarrow \quad \tilde{a}_t - rC_t &= \frac{\tilde{b}_t}{\sigma}(\mu - r) \\
\Rightarrow \quad \frac{\tilde{a}_t - rC_t}{\tilde{b}_t} &= \frac{\mu - r}{\sigma} \,.
\end{aligned}
$$

$$\tag{6.37}$$
$$\tag{6.38}$$

Finally, since $\tilde{a}_t = a_t C_t$ and $\tilde{b}_t = b_t C_t$, Eq. (6.38) becomes

$$\frac{a_t - r}{b_t} = \frac{\mu - r}{\sigma} \,.$$

This means that, as a consequence of the no-arbitrage condition, the market price of risk of the call C is equal to that of the underlying, namely the Sharpe ratios of C_t and S_t must coincide instantaneously.

6.2.3 Pricing equation for a call

Let us start now from Eq. (6.37), then we can write

$$\tilde{a}_t - \tilde{b}_t \frac{\mu - r}{\sigma} = rC_t \,.$$

$$\tag{6.39}$$

Furthermore, recalling that $\widetilde{a}_t = \dfrac{\partial C}{\partial t} + \mu S_t \dfrac{\partial C}{\partial S_t} + \dfrac{1}{2}\sigma^2 S_t^2 \dfrac{\partial^2 C}{\partial S_t^2}$ and $\widetilde{b} = \sigma S_t \dfrac{\partial C}{\partial S_t}$,

and substituting them in Eq. (6.39), we have

$$\frac{\partial C}{\partial t} + \frac{\partial C}{\partial S}\mu S_t + \frac{1}{2}\frac{\partial^2 C}{\partial S^2}\sigma^2 S_t^2 - \frac{\partial C}{\partial S}S_t(\mu - r) = rC_t$$

$$\Rightarrow \quad \frac{\partial C}{\partial t} + \frac{\partial C}{\partial S}rS_t + \frac{1}{2}\frac{\partial^2 C}{\partial S^2}\sigma^2 S_t^2 = rC_t, \qquad (6.40)$$

where there is not any term which depends from μ (see Remark 171 and Black and Scholes, 1973; Scholes, 1998). Thus rearranging Expression (6.40), we can write

$$\frac{1}{2}\sigma^2 S^2 \frac{\partial^2 C}{\partial S^2} + rS\frac{\partial C}{\partial S} + \frac{\partial C}{\partial t} = rC, \qquad (6.41)$$

which is the general equation of the Black-Scholes model for derivative pricing. Equation (6.41) is a (deterministic) second order partial differential equation of parabolic type, and describes the price of a call C_t, for $t \in [0, T]$.

Closed-form solution for European option pricing

Given the boundary conditions $C_T = \max(S_T - K, 0)$, one can demonstrate that the solution of Eq. (6.41) is

$$C_t = S_t \phi(d_1) - Ke^{-r(T-t)}\phi(d_2), \qquad (6.42)$$

where $d_1 = \dfrac{\log\left(\frac{S_t}{K}\right) + \left(r + \frac{\sigma^2}{2}\right)(T-t)}{\sigma\sqrt{T-t}}$ and $d_2 = d_1 - \sigma\sqrt{T-t}$, and $\phi(\cdot)$ is

the cumulative distribution function of a standard normal random variable (see Section 2.6.2).

Following the same steps and considerations used to find the call option price, given the boundary conditions $P_T = \max\{K - S_T, 0\}$, we can obtain the following closed-form solution for a European put option:

$$P_t = Ke^{-r(T-t)}\phi(-d_2) - S_t\phi(-d_1). \qquad (6.43)$$

Another way to achieve Expression (6.43) is to exploit the put-call-parity relationship $P_t = C_t - S_t + Ke^{-r(T-t)}$, and Eq. (6.42).

Exercise 174 (Black-Scholes parameters) *Solve the two following points.*

1. *Write a Function* F_BSPar *that provides the parameters d_1 and d_2 of the Black-Scholes formulas (6.42) and (6.43) as outputs, using the current price of the assets S_0, the rate of interest r, the volatilities*

---

of each asset σ, and the maturity of the options T as inputs. Then, write a Script S_BSPar1 which recalls the Function F_BSPar with the following inputs:

$$S_0 = \begin{bmatrix} 100 & 100 \\ 80 & 80 \end{bmatrix}, \quad K = \begin{bmatrix} 90 & 100 \\ 70 & 80 \end{bmatrix}, \quad \sigma = \begin{bmatrix} 0.2 & 0.2 \\ 0.3 & 0.3 \end{bmatrix},$$

$$T = \begin{bmatrix} 0.25 & 0.25 \\ 0.5 & 0.5 \end{bmatrix}, \quad and \quad r = 0.05.$$

Finally, using the Black-Scholes formulas (6.42) and (6.43), compute the price of the European call and put, respectively.

2. Let S be the $n \times 1$ vector of prices of a stock with a volatility σ. Furthermore, let us consider m options on this stock with different strike prices. Therefore, let K and T denote $1 \times m$ vectors of strike prices and maturities of the options, respectively. In another Script named S_BSPar2, using the Function F_BSPar, compute d_1 and d_2, considering the following parameters

$$S = \begin{bmatrix} 100 \\ 110 \\ 90 \end{bmatrix}, \quad K = \begin{bmatrix} 90 & 100 \end{bmatrix}, \quad \sigma = 0.4,$$

$$T = \begin{bmatrix} 0.25 & 0.25 \end{bmatrix}, \quad r = 0.05.$$

Then, using the Black-Scholes formulas (6.42) and (6.43), compute the price of the corresponding European call and put options. Note that the results must be $m \times n$ matrices. Hint: use the built-in function `repmat`.

Sol.: See Function F_BSPar and Scripts S_BSPar1, S_BSPar2.

Exercise 175 (Black-Scholes formulas) *Write a Script that computes the prices of European call and put options using the Black-Scholes formula. Consider the current price $S_0 = 100$, $\sigma = 0.2$, $r = 0.05$, $T = 1$, and $K = 110$. Use the Function F_BSPar of Exercise 174 to evaluate the parameters d_1 and d_2. Furthermore, to compute the cdf of a standard normal, use the the built-in function `erfc` (i.e., the complementary error function). For more details, see Remark 176.*

Sol.: See Script S_EurOpt_BS.

Remark 176 (Complementary error function) *The complementary error function can be defined as* $\mathrm{erfc}(x) = 1 - \mathrm{erf}(x)$. *Thus,*

$$\mathrm{erfc}(x) = \frac{2}{\sqrt{\pi}} \left(\int_0^{+\infty} e^{-t^2} dt - \int_0^x e^{-t^2} dt \right) = \frac{2}{\sqrt{\pi}} \int_x^{+\infty} e^{-t^2} dt. \quad (6.44)$$

It is straightforward to see that erfc(x) is an odd function, namely erfc($-x$) = $-$erfc(x). Hence, since erfc($-x$) = $1 -$ erf $(-x) = 1 +$ erf (x), we have

$$\phi(z) = \frac{1}{2}\left(1 + \text{erf}\left(\frac{z}{\sqrt{2}}\right)\right)$$
$$= \frac{1}{2}\text{erfc}\left(-\frac{z}{\sqrt{2}}\right).$$

Exercise 177 (Black-Scholes option pricing) *Write a Function that recalls* F_BSPar *and computes the price of European call and put options via the Black-Scholes formulas, considering the strike price K, the current price S_0, the maturity T, and the interest rate r as inputs, while the prices of the put and the call are the outputs. Then, write a Script that recalls the Function* F_BSPar *with $S_0 = 52.35$, $\sigma = 0.116$, $r = 0.035$, $T = 1$, and $K = 52.5$.*

Sol.: See Function F_OptionBS and Script S_OptionBS.

6.2.4 Implied volatility

In the real world, the option prices are generally determined by the market through the law of supply and demand. Thus, given the price of an option C_{market}, the maturity T, the strike price K, the current price of the underlying s_0 and the risk-free rate r, the only unknown parameter is the volatility σ, which is called *implied* volatility. Therefore, denoting by C the Black-Scholes formula (6.42) ad by σ_{impl} the implied volatility we have

$$C_{market} = C\left(s_0, T, \sigma_{impl}, r, K\right) . \tag{6.45}$$

Since when σ_{impl} rises, C increases, i.e., $\dfrac{\partial C}{\partial \sigma} > 0$, using (6.45) the implied volatility can be computed by means of numerical techniques, such as iterative methods based on the Newton algorithm. For more details on this topic, see, e.g., Hull and Basu (2016) and references therein. However, even though by assumption σ_{impl} should be constant for the Black-Scholes model (see Section 6.2.1), in the real market the implied volatility depends on T and K, thus showing the so-called volatility smile effect.

Remark 178 (Implied volatility surface) *Note that the structure of the implied volatility as a function of T and K, namely the implied volatility surface $\sigma_{impl}(K, T)$, can be used to calibrate the Black-Scholes model for pricing of Over The Counter derivative contracts, that are not quoted by the market.*

6.2.5 Black-Scholes formulas via integrals

Here we show how to obtain the closed-form solutions (6.42) and (6.43) of the Black-Scholes equation (6.41) for the evaluation of a derivative contract in terms of expected value under the risk-neutral probability measure $\mathbb{Q}$ of the future price of a call option, namely using the Feynman-Kac theorem. As mentioned in Remark 163, this theorem demonstrates that

$$C_t = e^{-r(T-t)} E_t^{\mathbb{Q}}(C_T) , \qquad (6.46)$$

where $E_t^{\mathbb{Q}}$ represents the expected value in t, i.e., conditioned by all information available until time t, under the risk-neutral probability measure $\mathbb{Q}$.

If the price of the underlying S is modeled by a geometric Brownian motion, then we explicitly know its cdf at time T. As briefly discussed in Section 6.1.2, under the subjective probability measure $\mathbb{P}$, we have

$$\text{under } \mathbb{P} \rightarrow S_T \sim \ln N\left(\ln S_t + \left(\mu - \frac{\sigma^2}{2}\right)(T-t), \sigma^2(T-t)\right) , \qquad (6.47)$$

while, under the risk-neutral probability measure $\mathbb{Q}$, the underlying S at time T has the following lognormal probability distribution

$$\text{under } \mathbb{Q} \rightarrow S_T \sim \ln N\left(\ln S_t + \left(r - \frac{\sigma^2}{2}\right)(T-t), \sigma^2(T-t)\right) . \qquad (6.48)$$

Given the boundary conditions $C_T = \max\{S_T - K, 0\}$ we can evaluate the price of a European call as follows

$$
\begin{aligned}
C_t &= e^{-r(T-t)} E_t^{\mathbb{Q}}[\max\{S_T - K, 0\}] \\
&= e^{-r(T-t)} E_t^{\mathbb{Q}}\left[(S_T - K)\mathbf{1}_{\{S_T > K\}}\right] \\
&= e^{-r(T-t)} E_t^{\mathbb{Q}}\left[S_T \mathbf{1}_{\{S_T > K\}}\right] - e^{-r(T-t)} K E_t^{\mathbb{Q}}\left[\mathbf{1}_{\{S_T > K\}}\right] , \qquad (6.49)
\end{aligned}
$$

where $\mathbf{1}$ is the indicator function. Let us analyze the second term of (6.49):

$$
\begin{aligned}
E_t^{\mathbb{Q}}\left[\mathbf{1}_{\{S_T > K\}}\right] &= \Pr^{\mathbb{Q}}(S_T > K) \\
&= \Pr^{\mathbb{Q}}\left(S_t e^{\left(r - \frac{\sigma^2}{2}\right) + \sigma(W_T - W_t)} > K\right) \\
&= \Pr^{\mathbb{Q}}\left(\ln S_t + \left(r - \frac{\sigma^2}{2}\right)(T-t) + \sigma\sqrt{T-t}\, Z_T > \ln K\right) .
\end{aligned}
$$

Thus, we have

$$E_t^{\mathbb{Q}}\left[\mathbf{1}_{\{S_T > K\}}\right] = \Pr^{\mathbb{Q}}\left(Z_T > \frac{\ln \frac{K}{S_t} - \left(r - \frac{\sigma^2}{2}\right)(T - t)}{\sigma\sqrt{T - t}}\right)$$

$$= \Pr^{\mathbb{Q}}\left(Z_T < \frac{\ln \frac{S_t}{K} + r(T - t)}{\sigma\sqrt{T - t}} - \frac{\sigma^2(T - t)}{2\sigma\sqrt{T - t}}\right)$$

$$= \Pr^{\mathbb{Q}}\left(Z_T < \frac{\ln \frac{S_t}{K} + r(T - t)}{\sigma\sqrt{T - t}} - \frac{\sigma\sqrt{T - t}}{2}\right).$$

Let us define $d_2 = \dfrac{\ln \frac{S_t}{K} + r(T - t)}{\sigma\sqrt{T - t}} - \dfrac{\sigma\sqrt{T - t}}{2}$. Hence, we can write

$$e^{-r(T-t)} K E_t^{\mathbb{Q}}\left[\mathbf{1}_{\{S_T > K\}}\right] = e^{-r(T-t)} K \Pr^{\mathbb{Q}}\left(Z_T < d_2\right)$$

$$= e^{-r(T-t)} K \phi(d_2), \qquad (6.50)$$

where $\phi(\cdot)$ is the cumulative distribution function of a standard normal random variable. We can now rearrange the first term of (6.49) by using the law of total expectation (see, e.g., Weiss et al, 2005), and by extracting every deterministic element from the expected value, given the information available at time t, as follows

$$e^{-r(T-t)} E_t^{\mathbb{Q}}\left[S_T \mathbf{1}_{\{S_T > K\}}\right] =$$

$$= e^{-r(T-t)} \left(E_t^{\mathbb{Q}}[S_T \mid S_T > K]\Pr^{\mathbb{Q}}(S_T > K) + 0 \cdot \Pr^{\mathbb{Q}}(S_T \leq K)\right)$$

$$= e^{-r(T-t)} E_t^{\mathbb{Q}}\left[S_t e^{\left(r - \frac{\sigma^2}{2}\right)(T-t) + \sigma\sqrt{T-t}Z_T} \mid S_T > K\right]\Pr^{\mathbb{Q}}(S_T > K)$$

$$= e^{-r(T-t) + r(T-t) - \frac{\sigma^2}{2}(T-t)} S_t E_t^{\mathbb{Q}}\left[e^{\sigma\sqrt{T-t}Z_T} \mid Z_T < d_2\right]\Pr^{\mathbb{Q}}(Z_T < d_2)$$

$$= e^{-\frac{\sigma^2}{2}(T-t)} S_t E_t^{\mathbb{Q}}\left[e^{\sigma\sqrt{T-t}Z_T} \mid Z_T > -d_2\right]\Pr^{\mathbb{Q}}(Z_T > -d_2)$$

$$= S_t \frac{1}{\sqrt{2\pi}} \int_{-d_2}^{+\infty} e^{-\frac{\sigma^2}{2}(T-t) + \sigma\sqrt{T-t}z} e^{-\frac{z^2}{2}} dz$$

$$= S_t \frac{1}{\sqrt{2\pi}} \int_{-d_2}^{+\infty} e^{-\frac{1}{2}\left(\sigma^2(T-t) - 2\sigma\sqrt{T-t}z + z^2\right)} dz$$

$$= S_t \frac{1}{\sqrt{2\pi}} \int_{-d_2}^{+\infty} e^{-\frac{(z - \sigma\sqrt{T-t})^2}{2}} dz.$$

Let us consider $-y = z - \sigma\sqrt{T-t}$, therefore $-dy = dz$ and the new limits of integration become $-\infty$ and $d_2 + \sigma\sqrt{T-t}$. Thus, we have

$$
\begin{aligned}
e^{-r(T-t)}E_t^{\mathbb{Q}}\left[S_T \mathbf{1}_{\{S_T > K\}}\right] &= -S_t \frac{1}{\sqrt{2\pi}} \int_{d_2+\sigma\sqrt{T-t}}^{-\infty} e^{-\frac{y^2}{2}}\,dy \\
&= S_t \frac{1}{\sqrt{2\pi}} \int_{-\infty}^{d_2+\sigma\sqrt{T-t}} e^{-\frac{y^2}{2}}\,dy \\
&= S_t \phi(\underbrace{d_2 + \sigma\sqrt{T-t}}_{=d_1}),
\end{aligned}
\tag{6.51}
$$

where we set $d_1 = d_2 + \sigma\sqrt{T-t} = \frac{\ln\frac{S_t}{K} + r(T-t)}{\sigma\sqrt{T-t}} + \frac{\sigma\sqrt{T-t}}{2}$. Now, substituting (6.50) and (6.51) in (6.49), we obtain the Black-Scholes formula for a European call pricing

$$
C_t = S_t\phi(d_1) - Ke^{-r(T-t)}\phi(d_2) .
\tag{6.52}
$$

Similarly, one can demonstrate that the price of a European Put is

$$
P_t = Ke^{-r(T-t)}\phi(-d_2) - S_t\phi(-d_1) .
\tag{6.53}
$$

6.3 Option Pricing via the Monte Carlo method

More in general, the Feynman-Kaĉ formula (6.46) holds for a generic derivative Y with payoff, at maturity T, $Y_T = f(S_T)$ (see, e.g., Fig. 6.8). Thus, the price of a generic derivative Y_t is

$$
Y_t = e^{-r(T-t)}E_t^{\mathbb{Q}}(Y_T) ,
\tag{6.54}
$$

where expectation is taken under the risk-neutral probability measure $\mathbb{Q}$. This means that if the price of the underlying S is represented by a geometric Brownian motion, then we know its explicit expression at time T (see Sections 5.2.3, 6.1.2, and 6.2.5), namely

$$
S_T = S_t e^{\left(r - \frac{\sigma^2}{2}\right)(T-t) + \sigma\sqrt{T-t}Z_T} .
\tag{6.55}
$$

Thus, Expression (6.54) also provides the logical scheme to compute the price of a derivative by means of the Monte Carlo method (see Section 5.1). This procedure consists in simulating, from t_0 to T, a large number of trajectories of the geometric Brownian motion, that at each time follows a log-normal distribution (6.48) under the risk-neutral probability measure $\mathbb{Q}$. More precisely, one has to

226

first discretize time into m intervals, $t_k \in [t_0, T]$ with $k = 0, 1, \cdots, m$, and then to generate n simulations for each time t_k

$$S_{t_k}^{(i)} = S_0 e^{\left(r - \frac{\sigma^2}{2}\right)t_k + \sigma W_{t_k}^{(i)}},$$

where $i = 1, 2, \cdots, n$. For instance, one can set $n = 10^6$. The second step is to compute the payoff Y at maturity T for each simulation, $Y_T^{(i)} = f\left(S_T^{(i)}\right)$ with $i = 1, 2, \cdots, n$ where, e.g., in the case of European call and put options it only depends on the realizations of S_T. If n is sufficiently large, then, according to the law of large numbers, the arithmetic mean of the simulations tends to the expected value

$$E_t^{\mathbb{Q}}[Y_T] = \frac{1}{n} \sum_{i=1}^{n} Y_T^{(i)}. \tag{6.56}$$

Hence, the price of an option with a generic payoff function $Y_T = f(S_T)$ can be computed as the discounted value of (6.56). In the following exercise, we apply the Monte Carlo methodology for pricing European call and put options. Note that if the number of simulations is sufficiently large, the Monte Carlo method, the Binomial model, and the Black-Scholes formulas have to lead to the same results, unless numerical errors.

Exercise 179 (Monte Carlo pricing) *Evaluate the price of a European call and put option using the Monte Carlo method. Recall that for a call we have*

$$C_0 = e^{-rT} E^{\mathbb{Q}}\left[\max\left(S_T - K, 0\right)\right],$$

while for a put

$$P_0 = e^{-rT} E^{\mathbb{Q}}\left[\max\left(K - S_T, 0\right)\right],$$

where the expected value is computed w.r.t. the risk-neutral probability measure $\mathbb{Q}$*. Under this probability measure, the asset price at maturity is*

$$S_T = S_0 \exp\left(\left(r - \frac{\sigma^2}{2}\right)T + \sigma\sqrt{T}Z\right),$$

where $Z \sim N(0, 1)$*. In other words, the drift coefficient of the geometric Brownian motion is equal to the risk-free-rate* r*.*
Write a Function to price a call and a put option via the Monte Carlo method. Then, write a Script for a call and a put pricing that recalls F_MonteCarlo, *considering a current price* $S_0 = 100$*, a diffusion coeffi-*

227

6.3.1 Path Dependent Derivatives

For a path dependent derivative, its payoff Y at maturity T is not only dependent on the underlying price S_T, but it is also a function of the prices that the underlying assumes during the life of the contract, namely between its starting date t_0 and its maturity T. More precisely, let $t_k \in [t_0, T]$, with $k = 0, 1, 2, \ldots, m$, then we have

$$Y(T) = f(S_{t_0}, S_{t_1}, S_{t_2}, \ldots, S_{t_m}) . \tag{6.57}$$

A typical example of path dependent derivatives is the one where the payoff at maturity is computed w.r.t. the average price of the underlying in the interval $[t_0, T]$. More precisely, considering, e.g., a call, its payoff at maturity T is

$$C_T = \max\{\bar{S} - K, 0\} ,$$

where $\bar{S} = \dfrac{1}{m+1} \displaystyle\sum_{k=0}^{m} S_{t_k}$. This kind of options are called *Asian* options.

Generally, the complexity of the payoff (6.57) does not allow to obtain a closed-form solution as in the case of European options. Thus, the path dependent derivative pricing is based on numerical procedures such as the Monte Carlo method, which is particularly suitable to this aim. Indeed, one can demonstrate that, also for the path dependent derivatives, the Feynman-Kac formula holds, namely

$$Y_{t_0} = e^{r(T-t_0)} E_{t_0}^{\mathbb{Q}} [Y_T] .$$

Thus, similarly to the case of European call and put options, we can consider the following steps for pricing Asian options:

1. discretize time into m intervals, $t_k \in [t_0, T]$ with $k = 0, 1, \cdots, m$;

2. generate a huge number n of trajectories, $S_{t_k}^{(i)} = S_0\, e^{\left(r - \frac{\sigma^2}{2}\right) t_k + \sigma W_{t_k}^{(i)}}$, where $i = 1, \ldots, n$;

3. compute the mean of each trajectory $\bar{S}^{(i)} = \frac{1}{m+1} \sum_{k=0}^{m} S_{t_k}^{(i)}$ for all $i = 1, \ldots, n$;

4. compute $E_{t_0}^{\mathbb{Q}} [Y_T] = \frac{1}{n} \sum_{i=1}^{n} \max\{\bar{S}^{(i)} - K, 0\}$;

5. compute the discounted value of $E_{t_0}^{\mathbb{Q}} [Y_T]$.

In the following exercise, we show how to apply this procedure in practice.

Exercise 180 (Path-dependent Option) *Evaluate the price of an Asian Option using the Monte Carlo method. More specifically, compute the price of an Asian call and put, where their payoffs at maturity T are respectively*

$$C_T^{(a)} = \max\left(\bar{S} - K, 0\right) \qquad and \qquad P_T^{(a)} = \max\left(K - \bar{S}, 0\right)$$

where $\bar{S}$ is the average price of the underlying S from today t_0 to the expiry T.

In a Script, price an asian call and put, considering the current price $S_0 = 36.64$, the historical volatility $\sigma = 0.2$, the risk-free rate $r = 0.0175$, the maturity $T = 0.5$, and the strike price $K = 37$. Furthermore, consider a number of simulations $n = 1000000$ and a number of time steps $m = 100$.

Sol.: See the Script S_MCAsianOp.

References

Acerbi C, Tasche D (2002) On the coherence of expected shortfall. Journal of Banking & Finance 26:1487–1503

Allais M (1984) The foundations of the theory of utility and risk some central points of the discussions at the oslo conference. In: Progress in utility and risk theory, Springer, pp 3–131

Artzner P, Delbaen F, Eber J, Heath D (1999) Coherent measures of risk. Mathematical Finance 9(3):203–228

Benati S, Rizzi R (2007) A mixed integer linear programming formulation of the optimal mean/value-at-risk portfolio problem. European Journal of Operational Research 176(1):423–434

Black F, Scholes M (1973) The pricing of options and corporate liabilities. Journal of Political Economy 81(3):637–654

Blay K, Markowitz H (2013) Risk-return analysis: The theory and practice of rational investing (volume one)

Brandimarte P (2013) Numerical methods in finance and economics: a MATLAB-based introduction. John Wiley & Sons

Buchanan JR (2012) An undergraduate introduction to financial mathematics. World Scientific Publishing Company

Carleo A, Cesarone F, Gheno A, Ricci JM (2017) Approximating exact expected utility via portfolio efficient frontiers. Decisions in Economics and Finance 40(1-2):115–143

Castagnoli E, Cigola M, Peccati L (2013) Financial Calculus with Application. EGEA

Castellani G, De Felice M, Moriconi F (2005a) Manuale di finanza. Vol. I. Tassi d'interesse. Mutui e obbligazioni. il Mulino

Castellani G, De Felice M, Moriconi F (2005b) Manuale di finanza. Vol. II. Teoria del portafoglio e del mercato azionario. il Mulino

Castellani G, De Felice M, Moriconi F (2006) Manuale di finanza. Vol. III. Modelli stocastici e contratti derivati. il Mulino

Cesarone F, Scozzari A, Tardella F (2013) A new method for mean-variance portfolio optimization with cardinality constraints. Annals of Operations Research 205(1):213–234

Cipra T (2010) Financial and insurance formulas. Springer Science & Business Media

Cont R (2001) Empirical properties of asset returns: stylized facts and statistical issues. Quantitative Finance 1(2):223–236

Cornuejols G, Tütüncü R (2006) Optimization methods in finance, vol 5. Cambridge University Press

Cox JC, Ross SA, Rubinstein M (1979) Option pricing: A simplified approach. Journal of Financial Economics 7(3):229–263

Elton EJ, Gruber MJ, Brown SJ, Goetzmann WN (2009) Modern portfolio theory and investment analysis. John Wiley & Sons

Fisher L, Weil RL (1971) Coping with the risk of interest-rate fluctuations: returns to bondholders from naive and optimal strategies. Journal of Business pp 408–431

Fong HG, Vasicek OA (1984) A risk minimizing strategy for portfolio immunization. The Journal of Finance 39(5):1541–1546

Gnedenko BV (2018) Theory of probability. Routledge

Hull JC, Basu S (2016) Options, futures, and other derivatives. Pearson Education India

Ingersoll JE (1987) Theory of financial decision making, vol 3. Rowman & Littlefield

Jensen JLWV (1906) Sur les fonctions convexes et les inégalités entre les valeurs moyennes. Acta mathematica 30(1):175–193

Konno H, Yamazaki H (1991) Mean-absolute deviation portfolio optimization model and its application to Tokyo stock exchange. Management Science 37:519–531

Levy H (2015) Stochastic dominance: Investment decision making under uncertainty. Springer

Lhabitant FS (2017) Portfolio Diversification. Elsevier

Markowitz HM (1952) Portfolio selection. Journal of Finance 7(1):77–91

Markowitz HM (1959) Portfolio selection: Efficient diversification of investments. Cowles Foundation for Research in Economics at Yale University, Monograph 16, John Wiley & Sons Inc., New York

Markowitz HM (2012) The "Great Confusion" concerning MPT. Aestimatio: The IEB International Journal of Finance (4):8–27

Markowitz HM (2014) Mean–variance approximations to expected utility. European Journal of Operational Research 234(2):346–355

Markowitz HM, Blay K (2013) Risk-Return Analysis: The Theory and Practice of Rational Investing (Volume One). McGraw Hill Professional

Merton RC (1973) Theory of rational option pricing. The Bell Journal of Economics and Management Science pp 141–183

Meucci A (2009a) Managing diversification. Risk 22:74–79

Meucci A (2009b) Risk and asset allocation. Springer, New York

Miettinen K (2012) Nonlinear multiobjective optimization, vol 12. Springer Science & Business Media

Mitra G, Kyriakis T, Lucas C, Pirbhad M (2003) A review of portfolio planning: Models and systems. In: Advances in portfolio construction and implementation, Elsevier, pp 1–39

Morgan J (1996) Riskmetrics-technical document. Tech. rep., New York: Morgan Guaranty Trust Company of New York, 4th ed.

Ogryczak W, Ruszczynski A (2002) Dual stochastic dominance and related mean-risk models. SIAM Journal on Optimization 13(1):60–78

Rachev ST, Hsu JS, Bagasheva BS, Fabozzi FJ (2008) Bayesian methods in finance, vol 153. John Wiley & Sons

Redington FM (1952) Review of the principles of life-office valuations. Journal of the Institute of Actuaries 78(3):286–340

Rockafellar RT, Uryasev S (2000) Optimization of Conditional Value-at-Risk. Journal of Risk 2:21–42

Scholes MS (1998) Derivatives in a dynamic environment. The American Economic Review 88(3):350–370

Shiu ES (1986) A generalization of Redington's theory of immunization. Actuarial Research Clearing House 2:69–81

Stoyanov SV, Rachev ST, Fabozzi FJ (2007) Optimal financial portfolios. Applied Mathematical Finance 14(5):401–436

Weiss N, Holmes P, Hardy M (2005) A Course in Probability. Pearson Addison Wesley, URL https://books.google.it/books?id=p-rwJAAACAAJ

Yitzhaki S (1982) Stochastic dominance, mean variance, and Gini's mean difference. The American Economic Review 72(1):178–185

Young MR (1998) A minimax portfolio selection rule with linear programming solution. Management Science 44(5):673–683

Suggested lesson plan

In this section, I propose a possible teaching timeline exploiting the content of this book. This approach has been tested during a long-time teaching experience at a Master's course in Finance to students with a relatively small background in programming, mathematics, probability and statistics.

Tables 6.1, 6.2 and 6.3 show an example of three modules that could be included in a Computational Finance course. I point out that each lesson should take around 100 minutes and that the proposed exercises could be performed live on a computer during the lessons or assigned as homework.

MODULE 1: Programming techniques for financial calculus		
Lesson	Contents	Lab Exercises
1	MATLAB basics; Preliminary elements	Ex. 12
2	Vectors and matrices; Standard operations of linear algebra; Element-by-element multiplication and division	Example 1-2 + Ex. 13
3	Colon (:) operator; Predefined and user-defined functions	Ex.14-15-16-17-18-19-21
4	M-file: Scripts and Functions	Example 3-4 + Ex. 33
5	Programming fundamentals: if, else, and elseif scheme	Example 5-6-7 + Ex. 22-23-26-28-29-30
6	Programming fundamentals: for loops	Example 8 + Ex. 27-34-35-37
7	Programming fundamentals: while loops	Example 9 + Ex. 36-41-42-43
8	MATLAB graphics	Example 10-11 + Ex. 38-39-40
9	portfolio of assets; term structure of spot and forward prices	Ex. 31-32-43-44-45-46-47-48-49
10	Net Present Value of assets; Amortization schedules	Ex. 50-51-52-53-54-55-56
11	Interest Rate Swap; Zero coupon swap curve; expected return and variance of a portfolio	Ex. 57-58-59-60

Table 6.1: Example of a lesson plan focused on programming techniques for financial calculus.

MODULE 2: Portfolio selection		
Lesson	**Contents**	**Lab Exercises**
12	Preliminary elements in Probability Theory and Statistics	Example 63-64-66-69-72-75-81-82
13	Random variables; Probability distributions	Example 88-93 + Ex. 95
14	Continuous random variables; probability distribution function; cumulative distribution function	Example 96 + Ex. 99
15	Higher-order moments and synthetic indices of a distribution; Uniform distribution	Example 101-102 + Ex. 100-103
16	Normal distribution	Example 105-106 + Ex. 107
17	Log-normal distribution; Chi-square distribution	Ex. 108-109
18	Student-t distribution; preliminary concepts on optimization; Linear Programming	Example 112 + Ex. 110
19	Quadratic Programming; Non-Linear Programming	Ex. 113-115
20	Multi-objective optimization; Efficient solutions and the efficient frontier; Portfolio of equities: Prices and returns	Ex. 117-119
21	Risk-return analysis; Mean-Variance model	Example 120 + Ex. 122-131-132
22	Capital Market Line; Market portfolio	Ex. 133
23	Effects of diversification for an Equally Weighted portfolio; Mean-Mean Absolute Deviation model	Ex. 134-135
24	Mean-Maximum Loss model; Value-at-Risk	Ex. 136-137
25	Mean-Conditional Value-at-Risk model; Mean-Gini model	Ex. 138-139
26	Bond portfolio immunization	Ex. 141-143-144-145

Table 6.2: Example of a lesson plan focused on portfolio selection.

MODULE 3: Derivatives pricing		
Lesson	**Contents**	**Lab Exercises**
27	Further elements on Probability Theory and Statistics; Introduction to Monte Carlo simulation; Stochastic process; Discrete random walk	Ex. 146
28	Wiener process; Brownian motion	Ex. 154
29	Ito's Lemma; Geometric Brownian motion	Example 160 + Ex. 161-162
30	Pricing of derivatives with an underlying security; Binomial model; A replicating portfolio of stocks and bonds	
31	Calibration of the binomial model; Multi-period case	Ex. 165-168
32	Black-Scholes model; Pricing of a European call: the dynamics of the option price; a hedging strategy	
33	Pricing equation for a call; Implied Volatility; Option Pricing via the Monte Carlo Method; Path Dependent Derivatives	Ex. 174-175-177-179-180

Table 6.3: Example of a lesson plan focused on derivatives pricing.